A Mathematical Gift, II

The interplay between
topology, functions,
geometry, and algebra

Mathematical World
Volume 20

A Mathematical Gift, II

The interplay between topology, functions, geometry, and algebra

Kenji Ueno
Koji Shiga
Shigeyuki Morita

Translated by Eiko Tyler

AMS
American Mathematical Society

KOKOSEI NI OKURU SUGAKU, 2

Kenji Ueno, Koji Shiga, Shigeyuki Morita

Originally published in Japanese in 1995 by Iwanami Shoten, Publishers, Tokyo, Japan

This English edition published by the American Mathematical Society, Providence, Rhode Island, by arrangement with the authors c/o Iwanami Shoten, Publishers, Tokyo

2000 *Mathematics Subject Classification.* Primary 00–01.

For additional information and updates on this book, visit
www.ams.org/bookpages/mawrld-20

Library of Congress Cataloging-in-Publication Data

Ueno, Kenji, 1945–
[Kokosei ni okuru sugaku II. English]
A mathematical gift II : the interplay between topology, functions, geometry, and algebra / Kenji Ueno, Koji Shiga, Shigeyuki Morita ; translated by Eiko Tyler.
p. cm. — (Mathematical world, ISSN 1055-9426 ; v. 20)
ISBN 0-8218-3283-2 (alk. paper)
1. Geometry. I. Shiga, Koji, 1930– II. Morita, S. (Shigeyuki), 1946– III. Title. IV. Series.

QA445.U3613 2004
516—dc22 2003062837

∞ The paper used in this book is acid-free and falls within the guidelines
established to ensure permanence and durability.
Visit the AMS home page at `http://www.ams.org/`

10 9 8 7 6 5 4 3 2 1 09 08 07 06 05 04

Contents

Preface

About two years ago, we developed a series of lectures on mathematics for high school students, which were held every other week at Kyoto University. Kenji Ueno, who is one of the authors of this book, began this program and was primarily responsible for initiating these lectures. This unique lecture format took root and grew due to Ueno's keen understanding of the education and learning process and the support and enthusiasm of approximately fifty high school and junior high students. Today you can find this mathematical lecture format not only at Kyoto University, but at many other universities in Japan.

On December 23 and 24, 1994, Kyoto University held a special program of lectures on Geometry as part of this series. The three authors of this book gave their presentations on different themes. After the scheduled lectures, two of the high school students who had attended asked if we have any plans to publish a book about the material we had just covered.

As it turns out, we were thinking about publishing a book based on this material because we felt that the lecture format could not reach enough people. Once we limit the presentation strictly to lectures, the audience drastically narrowed. Also, geographical limitations further reduce the number of people who have access to the information. So when we were asked if we would publish, we were determined to do so. At that time, though, we were not yet prepared to arrange and publish the lectures.

One night, one of the students suggested "Mathematics as a Gift to High School Students" as the title of the book. Initially, this suggestion surprised us. Inasmuch as our lectures to high school students were the same as those given to college students, we did not consciously differentiate them. The high school students had, however, picked up on a point that both we and our college students had missed—that education is a gift. It is a legacy that should be handed down and talked about through generations. I believe there is a similarity between the concept of learning and the concept of land as something to be inherited from those who came before us: something that can be handed down from generation to generation as a precious and irreplaceable object; I think we have lost sight of a very important philosophy that education is an invaluable gift. Something clicked in our minds that night. We decided to publish a book with the title the student suggested. The decision, which was made on the evening of December 24th, was a suitable one for Christmas Eve.

With the assistance of Iwanami-Shoten, our publisher, we are able to present this two-volume gift of mathematics. It is our hope that our readers will find the contents of these volumes to be like the joyous sound of a stream flowing from a "spring" called mathematics. We believe that this sound will be quite different from the mathematics you learned before. We believe that you will gain more than a simple joy by learning mathematics in a fun and comprehensive way. We hope you find the freedom of spirit and flexibility of mind that lie deep within rational thinking. We also hope that you will start reading this book with an attitude of optimism and finish it with a greater understanding of both mathematics and the human mind.

And now for a synopsis of the contents of this book: in the first chapter of Volume I,[1] we show that with the use of a limited number of mathematical equations, we can clarify the mysterious world of geometry hidden in curved surfaces. This clarification can be achieved by adopting a more global point of view in analyzing geometrical figures. This global point of view also will lead us towards the field called topology. This introduction to topology comes from the lectures of Shigeyuki Morita. At the end of Chapter 1, Lecture 2 is the proof of the Poincaré–Hopf theorem. The final step of this proof came from a student at one of Morita's lectures at the Tokyo Institute of Technology. This student approached Morita after the lecture and explained his proof method. Although Morita had shown a different method of proof in his lecture at the institute, he found the student's proof interesting and chose it for the book.

The second chapter addresses the topic of dimensions. You may find it interesting to know that we added this topic to the book as a result of many questions about the dimension asked at a gathering after one of our lectures. We have summarized these questions and presented them from a somewhat general point of view.

Chapter 3 in Volume II covers the journey down the road of the history of trigonometric functions. These functions were born from studying a circle. From a functional point of view they express the infinity hidden in circles. We finally reach the concept of power series expansion of functions. This road also leads to the birth of elliptic functions. This material comes from a lecture given by Koji Shiga.

The main theme of Chapter 4, which comes from the lecture of Kenji Ueno, is a geometric theorem called the Poncelet closure theorem. Among the many theorems of synthetic geometry developed during the nineteenth century, this theorem stands out for its extreme beauty of simplicity and depth. When you explore the relationship between ellipses and tangent lines which appear in the theorem from the point of view of algebraic geometry, you will arrive at the intersection point of geometry and algebra. These two distinctly different fields of mathematics influence, deepen, and expand upon

[1]Published by the American Mathematical Society as volume 19 of the series "Mathematical World" in 2003.

one another. This relationship continues to influence modern mathematics. Let us walk together down this road to a heightened understanding of the mystical properties hidden in geometrical figures. We hope that at the end of our journey you will experience the joy of learning geometry.

You can read volumes and chapters of this book in any order; they do not have to be studied in sequence. We hope you will accept and enjoy our gift.

March 1995

Koji Shiga

Chapter 3

The Legacy of Trigonometric Functions

Introduction

Mathematics is a unique discipline in that it deals with the intriguing concept of infinity. Fascinating as this may be to mathematicians, however, it may seem at first sight to have little relevance in the world of an average high school math student, whose tasks consist of proving formulas or finding solutions to problems. In fact, infinity plays a vital (if invidious) role in mathematics. More to the point, it is featured quite prominently in high school calculus. Calculus is the branch of mathematics which is based on real numbers and limits, both of which are intimately related to the notion of infinity. Therefore, one might say that calculus essentially is the study of infinity (even though when we try to find, say, the extrema of a cubic function using differentiation, we likely are not conscious of the fact that we are dealing with infinity). This lack of awareness has much to do with the fact that mathematics has been modified and has evolved over time into a tool that anyone can access. We now employ mathematics in virtually every aspect of science and technology. Paradoxically, it arguably is because of this omnipresence and "ease of access," as it were, that we may fail to realize that the concept of infinity is a foundation of the very "tools" with which we are so familiar.

In this chapter, we will discuss the history of a branch of mathematics which was developed by identifying the significance of a circle, a figure familiar to everyone. This branch of mathematics was developed around the 17th century, when calculus was still in its infancy.

When we see a full moon in the night sky, we cannot help but wondering at this mystical figure. Since ancient times, it has been suggested that part of that heavenly body's allure is its perfect shape (a disk). This captivating nature of the disk manifests itself *mathematically* as the ratio π of the length of a circle to its diameter. The length of half of the circle of radius one, is π. The peculiar allure of the circle is evidenced further by the centuries-long attempt to measure the exact length of the half-circle. Gradually, the value for π was approximated as 3.1, 3.14, 3.141, 3.1415, etc., as the values for the digits following the decimal point were further calculated. Eventually, people began to think that the journey to seek more values for the digits following the decimal point might, in fact, be endless. As the researchers continued to pursue their quest for the exact value of π, they eventually began to realize that the "end" of the journey was in the realm of infinity.

The seemingly endless sequence of numbers emanating from the "mystical" circle gave mankind the impetus to explore the concept of infinity.

Astronomers have long shown intense interest in the relationship between an arc and a chord of a circle. This relationship is expressed (in the study of functions) as the relationship between x and y in the sine function $y = \sin x$. The bridge between finite and infinite found in the relationship between the length of a chord and the length of an arc, in which the number π appears, is translated into the sine function $y = \sin x$. To study the properties of the sine function $y = \sin x$ in detail, it is necessary to introduce the concept of infinity. In turn, the concept of infinity gave rise to the concept of power expansion. Historically, mathematicians used the inverse sine function $y = \sin^{-1} x$, instead of the sine function $y = \sin x$, to study geometrically the relationship between a chord and an arc of a circle. Since the inverse sine function $y = \sin^{-1} x$ was easier to work with than the sine function $y = \sin x$, the infinite series for $y = \sin^{-1} x$ was obtained first. The successful use of the power series representation of $y = \sin^{-1} x$ was due to the effective application of **Newton's Generalized Binomial Theorem**. As you can see, trigonometric functions provided fertile soil to nourish seeds of mathematics.

Keep in mind that the functions we have just looked at deal with "perfect" circles. There also exist functions, developed by Gauss, Abel, and their contemporaries at the beginning of the 19th century, which deal with ellipses. These combined developments changed the mathematical landscape. With that background let us begin our discussion of trigonometric and elliptic functions.

LECTURE 1

Trigonometric Functions and Infinite Series

1.1. The Birth of Sine (sin), a Trigonometric Ratio

The phrase "Trigonometric Functions" may bring up images of right triangles (Figure 1.1) and trigonometric ratios such as sin, cos, and tan. Indeed, you likely learned about sin, cos, and tan as trigonometric ratios; later, you used them to study the relations between the angles and the sides of a triangle. Also, you probably studied certain formulas derived from addition theorems by regarding sin, cos, and tan as trigonometric functions. While you were studying trigonometric ratios or functions, you probably did not encounter anything spectacular. Despite that, it is not an exaggeration to say that trigonometric functions are a treasure chest of mathematics. Let us crack open the lid of this chest and take a peek at the glittering treasure inside.

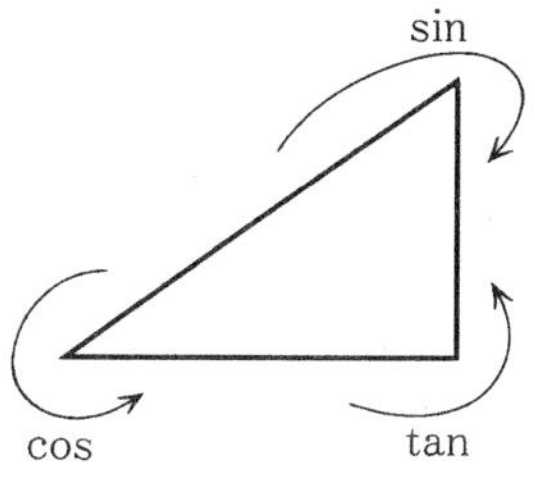

FIGURE 1.1

In the history of trigonometric functions **sine** (the trigonometric ratio) was the first to appear. This occurred about 2200 years ago. You might be surprised to know that the geometric figure which gave birth to sine was not the triangle but the circle. Hipparchus (approximately 190–125 B. C.), regarded by many as the father of Trigonometry, constructed a "table of chords" of a circle while he was attempting to verify numerically the results of many years of astronomical observations.

The "table of chords" of a circle is a table of the lengths of chords AA' corresponding to central angles α (see Figure 1.2). Since it is difficult to manually obtain the exact measurement of a chord, the length was approximated.

Hipparchus considered a circle with a diameter of 6875 units. From this circle, he obtained the lengths of chords corresponding to central angles

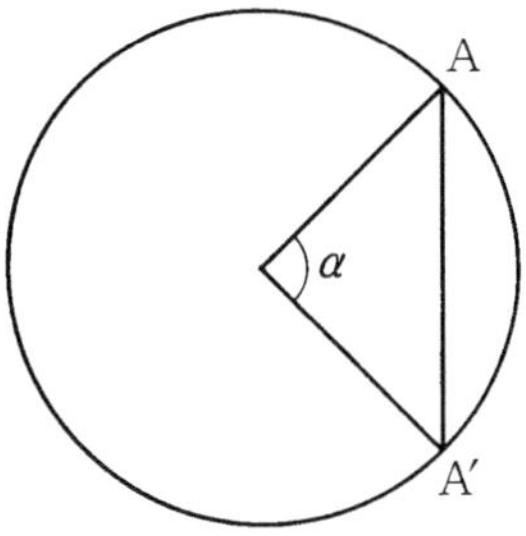

FIGURE 1.2

which are the integer multiples of 7.5 degrees. He then constructed the table of chords. Hipparchus' table of chords later was refined for additional accuracy and was a key element in the study of astronomical theory. Around 510 A. D., the Indian astronomer Aryabhata constructed a table consisting of half-chords. A table of half-chords is a table of values that represent the lengths of half-chords AB corresponding to the central angles α (see Figure 1.3).

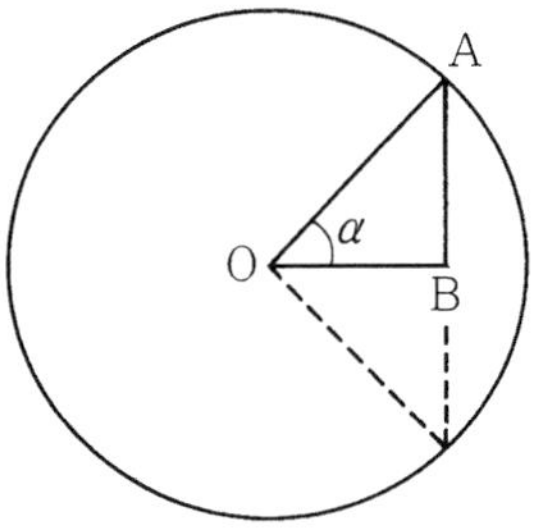

FIGURE 1.3

The "table of half-chords" is essentially the table of sines. Later we will explore the significance of the half-chord table, and discuss the relations between the degree measure and the radian measure of an angle.

In order to construct a table of half-chords, it is sufficient to use a single circle of a fixed radius. For example, if we construct a table of half-chords from a circle of radius 6 units, then according to the principle of similarity, the lengths of the corresponding half-chords of a circle of radius 12 units are twice those of a circle of radius 6. The lengths of the corresponding half-chords of a circle of radius 2 units are 1/3 unit. To simplify the usage and for standardization, let us construct a table of half-chords from a circle of radius 1 unit (the unit circle).

Next, we must determine how to measure the half of the central angle corresponding to a chord of a unit circle. If we refer to Figure 1.3, the measure of angle α corresponds to half-chord AB. The reader likely will recall measuring angles by the units of degrees with a protractor in grammar school. However, from the point of view of higher mathematics, the degree is not an approriate unit for measuring angles. One reason is that 180 degrees is the measure of the angle corresponding to half of a circumference. For example, if we divide a half-circumference into n arcs, then the corresponding central angle for each arc is $\frac{180}{n}$ degrees. The number 180 is an inconvenient obstacle in any calculation. Also, the number 180 is an arbitrary number and, speaking in mathematical terms, has no relation to the circumference of a circle (the number 180 is related to the sexagesimal system used by ancient Mesopotamians some 3000 years ago). As a new unit of angle measure, we should change 180° on the protractor to a half-circumference π of the unit circle and convert degree measure by multiplying by $\frac{\pi}{180^\circ}$ (see Figure 1.4). According to the conversion, 90° is $\frac{\pi}{2}$ and 60° is $\frac{\pi}{3}$. This improved method of measuring angles is called radian measure.

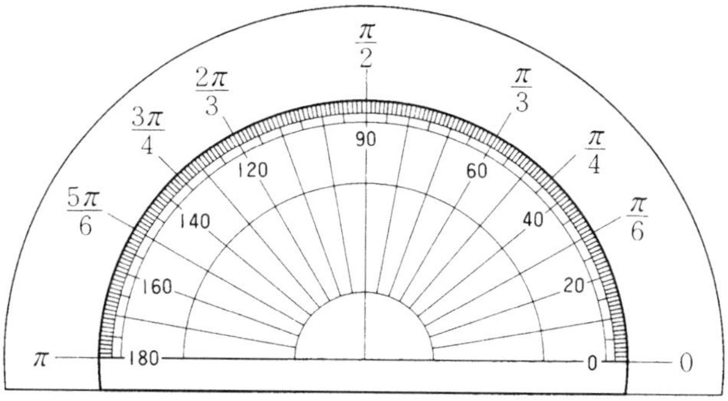

FIGURE 1.4

From now on, when we mention a circle, we will mean the unit circle (a circle of radius 1 unit). The measurement of angle α will be in radian measure. Therefore, if the central angle of the corresponding arc of the unit circle is α, then the length of the arc itself is also α (see Figure 1.5).

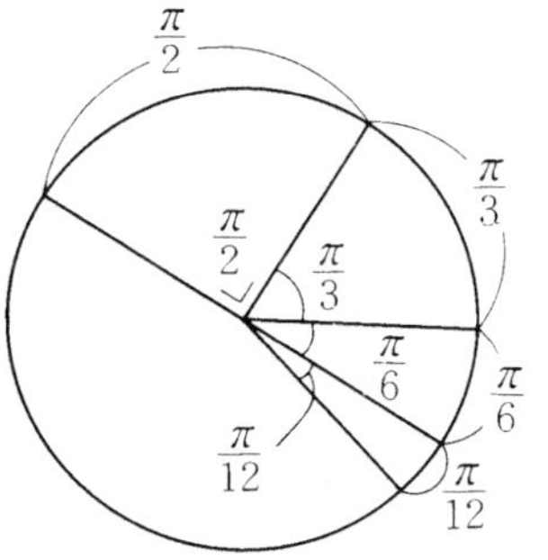

FIGURE 1.5

With this conversion of degree measure into radian measure, we now can consider the table of half-chords to be a table used to obtain the value y

(the length of a half-chord) when x is given (see Figure 1.6). This relation between y and x is the familiar sine function:

$$y = \sin x.$$

Therefore, a table of half-chords is the table of values of sine function.

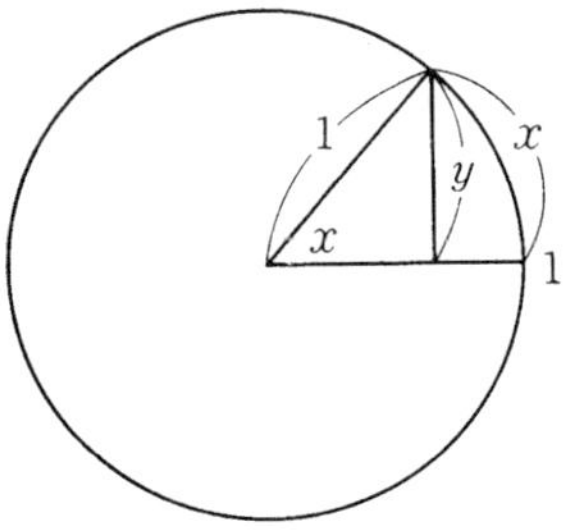

FIGURE 1.6

♣ Considering the relation between y and x in Figure 1.5, it is more suitable to call the trigonometric function $y = \sin x$ a "circular function". In fact, in recent years, the term "circular function" has sometimes replaced the term "trigonometric function" in mathematics textbooks. The definition of sine, cosine, and tangent as trigonometric ratios related to right triangles probably occurred in the 14th or 15th century, when Euclidean geometry was the central theme of mathematics.

1.2. The Trigonometric Function: $\sin x$

The trigonometric function $y = \sin x$ is the central theme of this section and we will restrict the angle x to the range $0 \le x < \frac{\pi}{2}$. In the expression of the function $y = \sin x$, x is a variable (precisely speaking, an independent variable). When x varies, the value of y varies accordingly. In that sense, y is a dependent variable. When the independent variable x is determined, then the dependent variable y is determined. From now on, we will regard the formula $y = \sin x$ as a functional relation between x and y, and we will study the properties of this relation.

The relation between x and y in the function $y = \sin x$ is given by the relation of central angles x, or the lengths of the corresponding arcs x, and the lengths of the corresponding half-chords y of the unit circle (see Figure 1.7, left). From this relation the additive formula

$$\sin(\alpha + \beta) = \sin\alpha\cos\beta + \sin\beta\cos\alpha$$

for sine is derived immediately using Figure 1.7.

CHALLENGE 1. Derive the additive formula for cos using similar arguments.

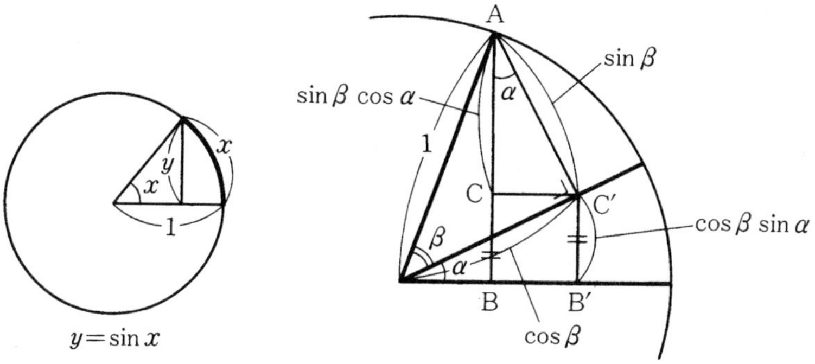

FIGURE 1.7

As you may have guessed by now, the main focus of the function sin is in the relation between the arc length x (measured in radians) and half-chord length y. However, the arc of a circle is curved and a half-chord is straight. We cannot measure a curved line with a straight edge ruler, nor can we measure a straight line with a curved measurement device. Curved lines and straight fines are like oil and water; they don't mix. For example, if we want to measure the length of an arc of a circle, we need to divide the arc into still finer arcs and measure the length of each chord of these finer arcs, then sum up the total measurements to obtain an approximation of the length of an arc of a circle (see Figure 1.8). Since ancient times, people tried to evaluate the half-circumference π using this approach.

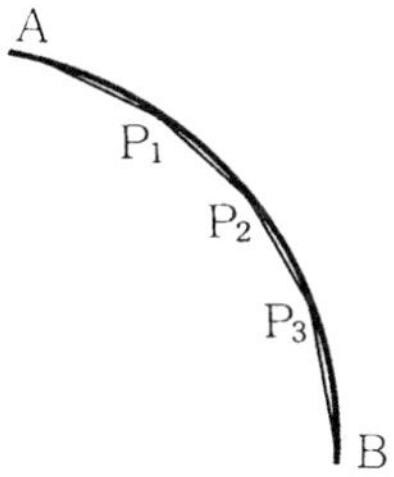

FIGURE 1.8

However, the half-circumference cannot be measured exactly. As you know, π is expressed as an infinite decimal fraction

$$\pi = 3.14159265358979323846\ldots.$$

The fact that π is an infinite decimal indicates that the half-circumference can never be measured by a finite number of operations. This shows that if we conceptualize a circle in terms of numbers, it will appear as "infinity".

♦ π is called an irrational number, i.e., a number that cannot be expressed as a fraction. Johann Heinrich Lambert (1728–1777) and Adrien Marie Legendre (1752–1833) proved this at the end of the 18th century. Therefore, if π is given as an infinite decimal fraction, it cannot be periodic.

The function $y = \sin x$ shows the relation between the arc lengths and the half-chord lengths in a circle. When we study this relation, we will encounter infinity. In fact, the importance of the function $y = \sin x$ is that the concept of infinity is hidden within the function. In studying $\sin x$ in detail, we will catch a glimpse of the shining source of mathematical creativity. This is why we earlier compared the collection of trigonometric functions to a treasure chest.

So far, we have been talking about the trigonometric function $y = \sin x$, with y being a dependent variable and x being an independent variable. Observe the two figures in Figure 1.9. Both of them show the relation between the arc length and half-chord length of a circle. Without thinking too deeply, which would you choose so that x will be an independent variable. In the lectures, a little less than 1/3 of the students who attended picked the left figure. The rest chose the figure on the right (although some of those students didn't really know why they chose thusly!).

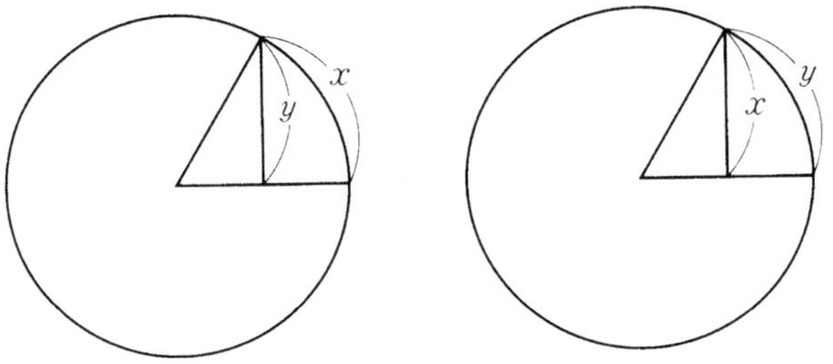

FIGURE 1.9

Whether by design or chance, the majority of the students have chosen the correct figure. When we study the relation between two variables, we normally choose x as the independent variable to represent a quantity which can be measured easily, and we choose y as the dependent variable to represent a quantity which is rather difficult to measure. For example, if we look at a train operation table, we will find that travel time, which is easy to measure, is given as the independent variable, and we will also find distance, which is not as easy to measure directly, as the dependent variable. However, if we consider an angular rotation or an equal division of an arc of a circle, then probably the figure on the left of Figure 1.9 would be a better choice. When we observe the figure on the right of Figure 1.9 in a geometric sense, the first thing we notice is the relation of lengths between the arc and the half-chord of the figure. It seems natural to choose the figure on the right, where x represents the length of a half-chord of a circle. Once we choose x to represent the length of a half-chord, it becomes possible to study geometrically the relation between x and y by pictorial representation. Doing just that, mathematicians in the 17th century interchanged the dependent and independent variables of the function $y = \sin x$, and a new avenue of research began.

The figure on the left of Figure 1.9 shows the relation between x and y as the familiar function

$$y = \sin x.$$

The figure on the right shows the relation between x and y as the function

$$y = \sin^{-1} x.$$

The function $y = \sin^{-1} x$ is called the **inverse sine function**, or the **arc sine function**. The relation between x and y is shown in Figure 1.10 as both a sine function and an inverse sine function.

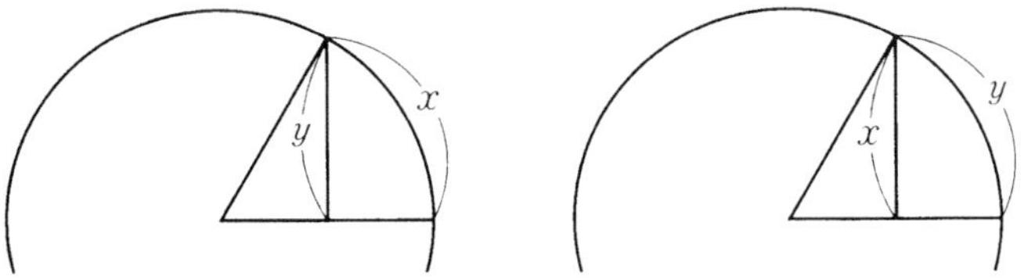

$y = \sin x$ $y = \sin^{-1} x$

FIGURE 1.10

1.3. $\sin^{-1} x$ and $\tan^{-1} x$

Now we will get to the heart of the quest for infinity (hidden in trigonometric functions) by studying the function $y = \sin^{-1} x$. On the way, we will discuss the function $y = \tan^{-1} x$ (the inverse tangent function). The path to infinity via $y = \tan^{-1} x$ presents different scenery than that via $y = \sin^{-1} x$. The function $y = \tan^{-1} x$ is the inverse function, which is obtained by reversing the relation between dependent and independent variables in the function $y = \tan x$. Figure 1.11 might help to explain the relation between dependent and independent variables of the sine and inverse sine functions.

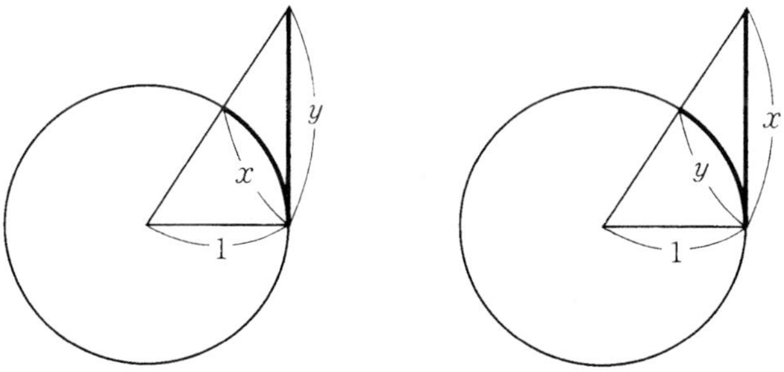

$y = \tan x$ $y = \tan^{-1} x$

FIGURE 1.11

Let us now observe the effect on the functions $y = \sin^{-1} x$ and $y = \tan^{-1} x$ when we make small changes to the value of x. Let us change the independent variable x by a small amount and then observe the corresponding change in the dependent variable y. The change of x is denoted by Δx, and is called an **increment of** x. The corresponding change of y is denoted by Δy. Our goal is to approximate Δy in terms of Δx, that is, to find an approximate relation between Δx and Δy. We can best understand this relation by studying the geometric representation for Δx and Δy corresponding to the functions $y = \sin^{-1} x$ and $y = \tan^{-1} x$ (see Figure 1.12).

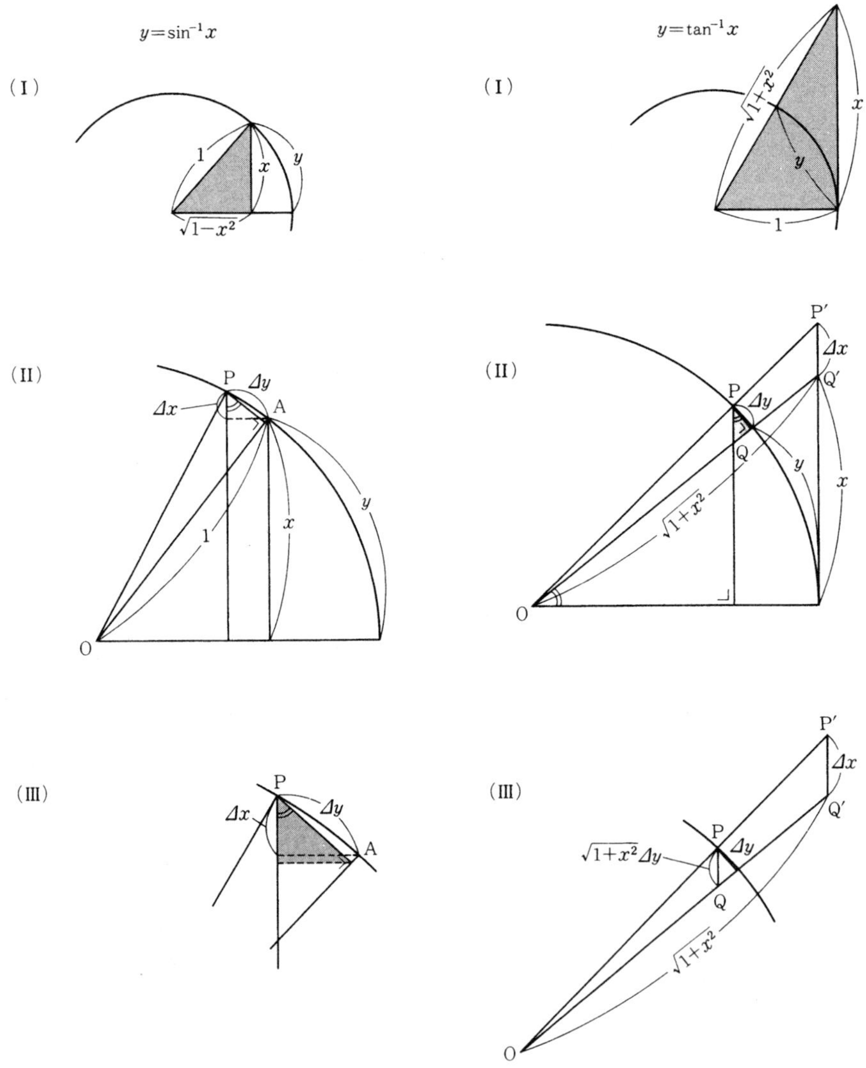

FIGURE 1.12

In Figure 1.12, the left column refers to the function $\sin^{-1}(x)$, and the right column refers to the function $\tan^{-1}(x)$. For the function $y = \sin^{-1}(x)$, relations in picture (I) hold by the Pythagorean theorem. In picture (II), the increment Δx and the corresponding increment Δy are shown. The increase $\widehat{AP}$ is shown in picture (III). In the shaded triangle the hypotenuse is $\approx \Delta y$ and the base is $\approx \Delta x$. This triangle is similar to the shaded triangle in picture (I). Therefore $\frac{\Delta y}{\Delta x} \approx \frac{1}{\sqrt{1-x^2}}$, hence $\Delta y \approx \frac{1}{\sqrt{1-x^2}}$.

For the function $y = \tan^{-1} x$ relations in picture (I) hold by the Pithagorean theorem. In picture (II), the increment Δx and the corresponding increase Δy are shown. The small shaded triangle is similar to the shaded triangle in picture (I). Hence, $PQ \approx \sqrt{1+x^2}\,\Delta y$. In picture (III), the triangle OPQ is similar to the triangle $OP'Q'$. For small Δx, $OQ \approx 1$. Hence, $\frac{\sqrt{1+x^2}\,\Delta y}{1} \approx \frac{\Delta x}{\sqrt{1+x^2}}$. Therefore, $\Delta y \approx \frac{1}{1+x^2}\Delta x$.

Hence, the relation between Δx and Δy is given as follows:

$$\text{If the function is } y = \sin^{-1} x, \text{ then } \Delta y \approx \frac{1}{\sqrt{1-x^2}}\Delta x.$$

$$\text{If the function is } y = \tan^{-1} x, \text{ then } \Delta y \approx \frac{1}{1+x^2}\Delta x.$$

The two approximate expressions look differently due to the difference in the domains in which the independent variable x varies. In these expressions, as the increment Δx of the independent variable tends to zero, the approximation of the increment Δy of the dependent variable approaches the precise value. For both functions related to Figure 1.12 we obtain an approximate expression by estimating the length of a very small arc of a circle in terms of the line segment PR in Figure 1.13.

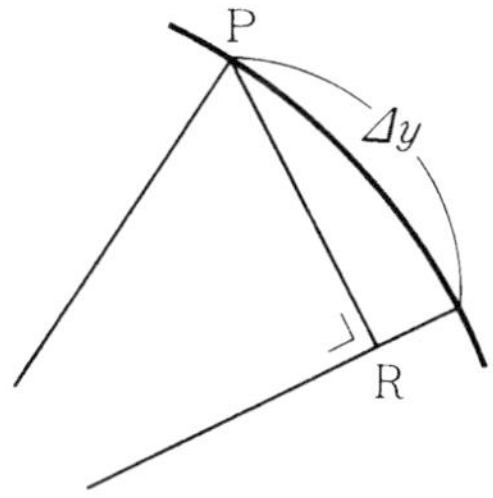

FIGURE 1.13

1.4. Integral Representations and Power Series Representations of Functions

If the independent variable x undergoes a small change Δx, then the corresponding change Δy in the dependent variable y is approximated as follows:

(1) if the function is $y = \sin^{-1} x$, then $\Delta y \approx \dfrac{1}{\sqrt{1-x^2}}\,\Delta x$;

(2) if the function is $y = \tan^{-1} x$, then $\Delta y \approx \dfrac{1}{1+x^2}\,\Delta x$.

(For the following paragraph, refer to Figure 1.14.)

Consider an arc with fixed length y. Taking n to be any positive integer, let us divide the arc equally into n smaller arcs by choosing points P_0, P_1, P_2, ..., P_n, where $P_0 = E$ and P_n are the endpoints of the arc and P_k are the dividing points for $k = 1, 2, \ldots, n-1$. We then estimate the length of each arc $\widehat{P_kP_{k+1}}$ by using the above approximate expressions. Let R_1, R_2, ..., R_n be the intersection points of the lines passing through P_1, P_2, ..., P_n and perpendicular to the radii OE, OP_1, ..., OP_{n-1}, respectively. Then the length of each perpendicular segment P_kR_k is approximated by Δx (see Figure 1.14). Hence the length of the arc $\widehat{EP_n}$ is approximated as follows:

$$
\begin{aligned}
y &= (\text{length of } \widehat{EP_1}) + (\text{length of } \widehat{P_1P_2}) + \cdots + (\text{length of } \widehat{P_1P_2}) \\
&\approx (\text{length of } P_1R_1) + (\text{length of } P_2R_2) + \cdots + (\text{length of } P_nR_n).
\end{aligned} \tag{3}
$$

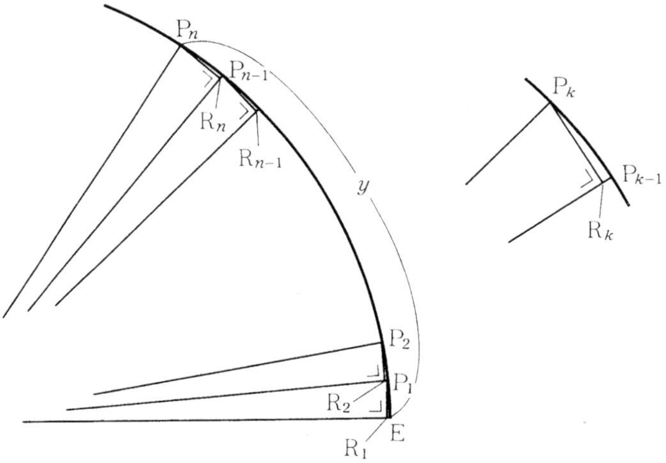

FIGURE 1.14

We have arrived at the approximate expression (3) by studying the preceding geometric figures. But to refine this expression, we must divide the arc into smaller and smaller arcs. Then the total length of the perpendicular segments gets closer and closer to the exact length of the arc. To replace the approximation with an equality, we must briefly depart the finite world by letting the integer n approach infinity, thus crossing over into the world of "infinity". As mentioned previously, to clarify the relation between x and y, we must walk the path to infinity. As the integer n tends to infinity, which is written as $n \to \infty$, the limit of the total length of the perpendicular segments is given by the definite integral as follows:

$$(4) \qquad y = \int_0^x \frac{dt}{\sqrt{1-t^2}} \quad \text{(obtained by using expression (1) above),}$$

$$(5) \qquad y = \int_0^x \frac{dt}{1+t^2} \quad \text{(obtained by using expression (2) above).}$$

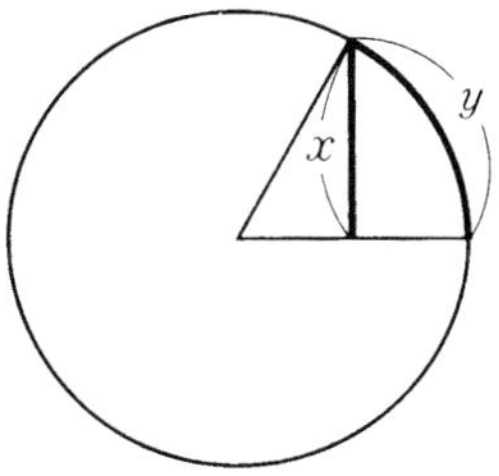

FIGURE 1.15. The relation between x and y in equation (4).

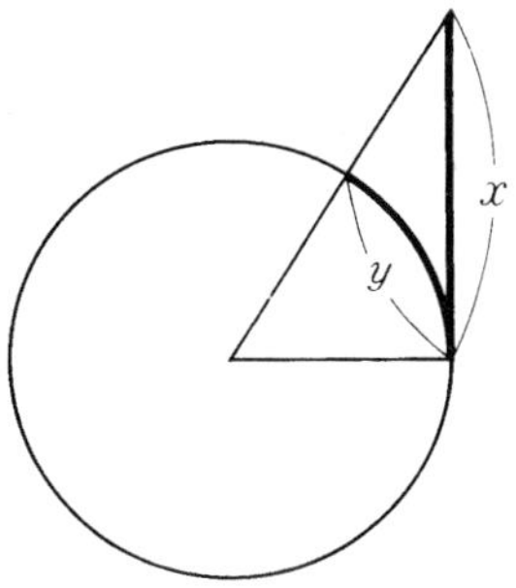

FIGURE 1.16. The relation between x and y in equation (5).

♣ For those of you who have not studied integral calculus, let us briefly discuss integrals. The notion of integral originated in the search for the areas of geometric figures. If a nonnegative function $y = f(x)$ is defined for real numbers between a and b, then we obtain the area under the graph of $y = f(x)$ from a to b using the definite integral. The area of the region bounded above by the graph of $y = f(x)$, below by the x-axis, and by vertical lines $x = a$ and $x = b$ is called the definite integral of $y = f(x)$ from a to b. It is denoted by

$$(*) \qquad \int_a^b f(x)\,dx.$$

The numbers a and b are called the **limits of integration**.

To obtain the definite integral of $y = f(x)$ from a to b (the area under the graph of $y = f(x)$), let us divide the interval $[a, b]$ into n equal subintervals by choosing numbers $x_0, x_1, \ldots, x_n$ such that $a = x_0 < x_1 < x_2 < \cdots < x_n = b$ and $x_k - x_{k-1} = \frac{b-a}{n}$. If we denote $\frac{b-a}{n}$ by Δx, then the area under the graph of $y = f(x)$ from a to b is approximated by the sum of the areas of the n rectangles (shaded figures in Figure 1.17) as follows:

$$(**) \qquad f(x_0)\Delta x + f(x_1)\Delta x + \cdots + f(x_{n-1})\Delta x.$$

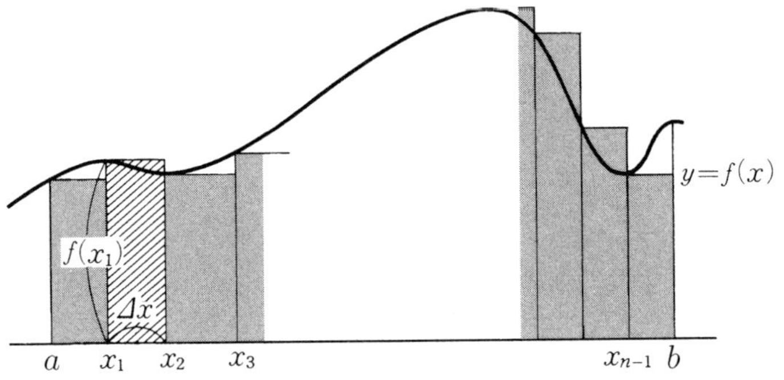

FIGURE 1.17. The area of the distinguished rectangle equals $f(x_1)\Delta x$.

If we let Δx approach zero ($\Delta x \to 0$) by further dividing the interval, then the sum $(**)$ approaches the area of the region below the graph of $y = f(x)$ from a to b. The definite integral $\int_a^b f(x)\,dx$ is, in a way, defined as the limit of the sum $(**)$.

Once we established the notion of the integral using this limit process, we can apply the limit process to any sum of the form $(**)$, even if the sum does not represent the approximation of the area. In fact, the definite integral of a function $y = f(x)$ from a to b is defined precisely as the limit of the so-called Riemann sums of the function. The area under the graph of $y = f(x)$ from a to b is given by the definite integral $(*)$ according to this precise definition.

In expressions (4) and (5), the limits of integration are zero and x. The expressions $\frac{1}{\sqrt{1-t^2}}$ and $\frac{1}{1+t^2}$ under the integral signs (called the integrands) came from $\frac{1}{\sqrt{1-x^2}}$ and $\frac{1}{1+x^2}$, respectively, by replacing x with t.

The expressions (4) and (5) can be rewritten as follows:

$$(6) \qquad y = \sin^{-1} x = \int_0^x \frac{1}{\sqrt{1-t^2}}\,dt,$$

$$(7) \qquad y = \tan^{-1} x = \int_0^x \frac{1}{1+t^2}\,dt.$$

We will now discuss the integral representations of the above two functions. The expressions (4) and (5) show that the "mother" of $\sin^{-1} x$ is $\frac{1}{\sqrt{1-t^2}}$ and that of $\tan^{-1} x$ is $\frac{1}{1+t^2}$. Referring to Figure 1.12, we see that the expression $\frac{1}{\sqrt{1-t^2}}$ is derived directly from the Pythagorean theorem. But the expression $\frac{1}{1+t^2}$ is derived rather indirectly because the relation between x and y in the function $y = \tan^{-1} x$ with respect to a unit circle is indirect, resulting in a form other than the Pythagorean theorem. But if we restrict ourselves to observing the end results of our geometric proofs, the equation (7) has a much simpler form than that of equation (6).

However, in the case of equation (6), the arc length y and the half-chord length x must be related to each other via an infinite expression, as the formula $\pi = 3.141592\ldots$ shows, and that infinite expression must somehow be incorporated in the integral notation on the right-hand side of equation (6). (At this point, the reader may feel that although the entrance door to the world of infinity is opening, yet the interior doors seem to remain tightly shut!)

We will see how x and y are related to each other by removing the integral signs from equations (6) and (7). At first glance, it appears that it will be easier to start with equation (7). The integrand is given by the infinite geometric series[1] as follows:

$$\frac{1}{1+t^2} = 1 - t^2 + t^4 - t^6 + t^8 - \cdots \qquad (|t| < 1).$$

This equation shows that if the degree n of a polynomial increases to infinity, the resulting infinite series represents the fractional expression. Fortunately, since the expression $\frac{1}{1+t^2}$ is under the integral sign of the definite integral (7), the integration is rather simple. By utilizing the integral formula

$$\int_0^x t^n \, dt = \frac{1}{n+1} x^{n+1},$$

[1]The partial sum S_n of the geometric series is given by

$$(\#) \qquad a + ar + \cdots + ar^{n-1} = \frac{a(1-r^2)}{1-r} \quad \text{for } r \neq 1.$$

We obtain this formula by comparing S_n with the left-hand side of the equation and subtracting rS_n from S_n. We obtain $S_n - rS_n = (1-r)S_n = a - a^n$. Therefore $S_n = \frac{a(1-r^2)}{1-r}$. The sum of the infinite geometric series is given by $a + ar + \cdots + ar^n + \cdots = \frac{a}{1-r}$. We obtain this formula by applying the limit process to S_n. To accomplish this, we first rewrite formula (#) as $a + ar + \cdots + ar^{n-1} = \frac{a}{1-r} - \frac{ar^2}{1-r}$. The limit of S_n as n tends to infinity is equal to the limit of $\left(\frac{a}{1-r} - \frac{ar^2}{1-r}\right)$. If $|r| < 1$, then the limit of r^n as n tends to infinity is zero ($r^n \to 0$ as $n \to \infty$). Therefore, $a + ar + \cdots + ar^n + \cdots = \frac{a}{1-r}$.

and by using the termwise integration, we have

$$\begin{aligned} y = \tan^{-1} x &= \int_0^x \frac{1}{1+t^2}\,dt \\ &= \int_0^x (1 - t^2 + t^4 - t^6 + t^8 - \cdots)\,dt \\ &= x - \frac{1}{3}x^3 + \frac{1}{5}x^5 - \frac{1}{7}x^7 + \frac{1}{9}x^9 - \cdots \quad (|x| < 1). \end{aligned}$$

When the "veil" of the integral sign is removed, as it were, from the definite integral of the function $y = \tan^{-1} x$, the relation between x and y is revealed clearly in the infinite series representation of y. This fact is stated as follows:

$$(8) \qquad y = \tan^{-1} x = x - \frac{1}{3}x^3 + \frac{1}{5}x^5 - \frac{1}{7}x^7 + \frac{1}{9}x^9 - \cdots.$$

The right-hand side of the above equation is called the **Power Series Expansion** of the function $y = \tan^{-1} x$. Let us approximate $y = \tan^{-1} x$ by using the first five terms of the power series expansion for $x = 0.8$. The computation gives y as approximately 0.6798. Since $\tan^{-1} 0.8 = 0.67474\ldots$, the error is less than $\frac{5}{1000}$ (see Figure 1.18).

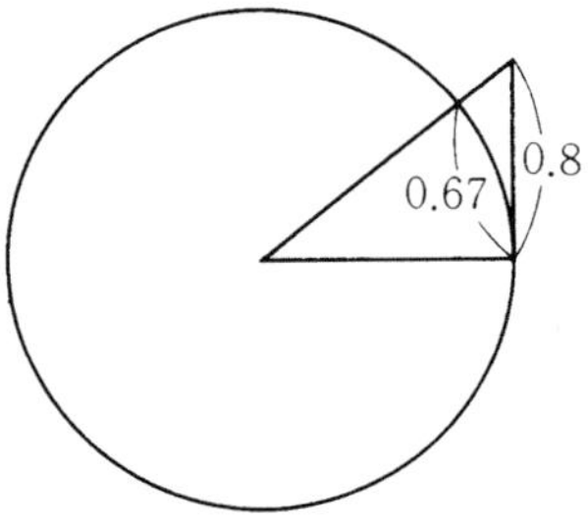

FIGURE 1.18

It has been shown that the right-hand side of (8) will converge even if $x = 1$, and we find the sum, in this case, to be $\frac{\pi}{4}$. If $x = 1$, this power series becomes the famous **Leibniz Series** for $\frac{\pi}{4}$, namely

$$\frac{\pi}{4} = 1 - \frac{1}{3} + \frac{1}{5} - \frac{1}{7} + \frac{1}{9} - \cdots.$$

The power series expansion for the function $y = \tan^{-1} x$ was known to mathematicians in India as early as the 15th century. In Europe, this series remained unknown till the mid-17th century, when it was discovered by Nicolaus Mercator (1620–1687) and John Wallis (1616–1703). In contrast to the function $y = \tan^{-1} x$, it is not easy to represent the function $y = \sin^{-1} x$

by a power series. The reason that it is not easy is that the integrand

$$\frac{1}{\sqrt{1-t^2}} \tag{9}$$

was not well understood. So let us now study this expression. The first thing we note is that it cannot be changed into any other form (e.g., infinite series).

Recall that when we discussed the function $y = \tan^{-1} x$, we found that the sum of the infinite geometric series played a decisive role in deriving expression (8). Therefore, for the function $y = \sin^{-1} x$ there must also be some important clue hidden in expression (9), which will yield the infinite power series. This is indeed the case, as Newton was first to discover.

When Newton was in his early twenties (in 1664 or 1665), he studied the **General Binomial Theorem** and obtained the following formula:

> If α is a positive or negative rational number, then
>
> $$(1+x)^\alpha = 1 + \frac{\alpha}{1!}x + \frac{\alpha(\alpha-1)}{2!}x^2 + \frac{\alpha(\alpha-1)(\alpha-2)}{3!}x^3 + \cdots + \frac{\alpha(\alpha-1)(\alpha-2)\cdots(\alpha-k+1)}{k!}x^k + \cdots \quad \text{for } |x| < 1.$$

The notation $k!$ stands for k factorial.[2]

CHALLENGE 2. Show that

$$(1+x)^2 = 1 + 2x + x^2,$$
$$(1+x)^3 = 1 + 3x + 3x^2 + x^3.$$

CHALLENGE 3. Using the binomial theorem, find the power series expansion for $(1+x)^6$.

---

[2]The value obtained by multiplying all the natural numbers from 1 to n is denoted by $n!$. Some examples are shown below:

$$\begin{aligned}
1! &= 1,\\
2! &= 2 \cdot 1 = 2,\\
3! &= 3 \cdot 2 \cdot 1 = 6,\\
4! &= 4 \cdot 3 \cdot 2 \cdot 1 = 24,\\
5! &= 120,\\
6! &= 720,\\
7! &= 5040,\\
8! &= 40320,\\
9! &= 362880,\\
10! &= 3628800.
\end{aligned}$$

CHALLENGE 4. Show that in order that the power series expansion for $(1+x)^{\alpha}$ be a polynomial (a finite series), α must be either a natural number or zero.

CHALLENGE 5. If $\alpha = -1$, show that $(1+x)^{\alpha}$ is equal to the sum of a geometric series.

CHALLENGE 6. If $\alpha = \frac{1}{2}$, approximate $\sqrt{1+x}$ by finding the first five terms of the infinite power series expansion.[3]

According to the general binomial theorem, if $\alpha = -\frac{1}{2}$ and $x = -t^2$, then

$$\frac{1}{\sqrt{1-t^2}} = (1-t^2)^{-\frac{1}{2}} = 1 + \frac{1}{2}t^2 + \frac{1\cdot 3}{2\cdot 4}t^4 + \frac{1\cdot 3\cdot 5}{2\cdot 4\cdot 6}t^6 + \cdots.$$

Thus, it is clear that if we remove the integral notation from expression (6), then the reiationship between x and y is as follows (with y represented by the infinite series):

$$\begin{aligned} y = \sin^{-1} x &= \int_0^x \frac{1}{\sqrt{1-t^2}}\, dt \\ &= \int_0^x \left(1 + \frac{1}{2}t^2 + \frac{1\cdot 3}{2\cdot 4}t^4 + \frac{1\cdot 3\cdot 5}{2\cdot 4\cdot 6} + \cdots\right) dt \\ &= x + \frac{1}{2}\frac{x^3}{3} + \frac{1\cdot 3}{2\cdot 4}\frac{x^5}{5} + \frac{1\cdot 3\cdot 5}{2\cdot 4\cdot 6}\frac{x^7}{7} + \cdots. \end{aligned}$$

The above fact is restated as

(10) if $|x| < 1$, then $y = \sin^{-1} x = x + \frac{1}{2}\frac{x^3}{3} + \frac{1\cdot 3}{2\cdot 4}\frac{x^5}{5} + \frac{1\cdot 3\cdot 5}{2\cdot 4\cdot 6}\frac{x^7}{7} + \cdots.$

The right-hand side of this equation is called the infinite power series expansion of $y = \sin^{-1} x$. Pay attention to the coefficients of the terms of the series. For example, the coefficient of x^9, following x^7, is given by

$$\frac{1\cdot 3\cdot 5\cdot 7}{2\cdot 4\cdot 6\cdot 8}\cdot\frac{x^9}{9}.$$

As you can see, even numbers and odd numbers appear in an orderly sequence.

We tend to think that a circle has nothing to do with the world of numbers. Returning to our earlier analogy, this thought is reinforced when we look at the full moon. Certainty, we do not think of numbers when we gaze upon the moon's mysterious beauty. But, in fact, as in expression (10), even numbers and odd numbers appear in the relation between the arc length and the half-chord length of a circle. This fact is intriguing. No one knows

[3]The definition of $a^{\frac{n}{m}}$ and a^{-n} can be given as follows: if both m and n are natural numbers, then $a^{\frac{m}{n}} = \sqrt[m]{a^n}$ for $a > 0$. If n is a natural number, then $a^{-n} = \frac{1}{a^n}$ for $a > 0$.

why such a harmonious relation exists between a circle and the world of numbers. Mathematicians know only that equation (10) is valid, but they cannot explain the reason behind it.

1.5. Reversing Infinite Power Series

So far, by working through the process of deriving infinite series from a circle, we have seen the power series expansions of the functions $y = \tan^{-1} x$ and $y = \sin^{-1} x$. However, geometry not withstanding, the most important trigonometric function for the development of mathematics in the various fields of science is $y = \sin x$. Can the function $y = \sin x$ be represented by a power series expansion as

$$y = \sin x = a_0 + a_1 x + a_2 x^2 + a_3 x^3 + \cdots + a_n x^n + \cdots \tag{11}$$

for suitable coefficients $a_0, a_1, a_2, \ldots$?

If it is possible to represent this function by the power series expansion, then what are the coefficients $a_0, a_1, a_2, \ldots$?

♣ If you studied calculus, then you know that $(\sin x)' = \cos x$. Hence, we know that $\sin x = \int_0^t \cos t \, dt$. In this integral representation of $\sin x$, $\cos t$ follows the integral notation. But in the integral representation of $\sin^{-1} x$, $\frac{1}{\sqrt{1-t^2}}$ follows the integral notation. Using the **Binomial Theorem** and the **Termwise Integration Theorem**, $\sin^{-1} x$ is represented by the power series expansion. We cannot apply the same method to $\cos t$ (which follows the integral notation) to find the power series expansion of $\sin x$.

Recognizing this, Isaac Newton instead employed a reverse method to obtain the power series. To use Newton's method, we must first assume that the function $y = \sin x$ is represented by the power series in expression (11) for suitable $a_0, a_1, a_2, \ldots$. In this function, x represents the arc length and y represents the half-chord length. The relation between the two functions sine and arcsine is given by the following diagram:

$$\text{the arc length} \xrightarrow{\sin} \text{the half-chord length}$$

$$(x) \xleftarrow[\sin^{-1}]{\qquad\qquad} (y)$$

This relation is simply stated as

$$y = \sin x \Leftrightarrow x = \sin^{-1} y.^{4}$$

Newton attempted to use the following method to find the coefficients $a_0, a_1, a_2, \ldots$.

[4] The notation ($\Leftrightarrow$) is read "if and only if" and means equivalence.

First, he represented $x = \sin^{-1} y$ by the power series as

$$x = y + \frac{1}{2}\frac{y^3}{3} + \frac{1\cdot 3}{2\cdot 4}\frac{y^5}{5} + \frac{1\cdot 3\cdot 5}{2\cdot 4\cdot 6}\frac{y^7}{7} + \cdots .$$

Substituting the sum of the power series in (11) for y we obtain

$$\begin{aligned} x &= (a_0 + a_1x + a_2x^2 + \cdots + a_nx^n + \cdots) \\ &+ \frac{1}{2\cdot 3}(a_0 + a_1x + a_2x^2 + \cdots + a_nx^n + \cdots)^3 \\ &+ \frac{1\cdot 3}{2\cdot 4\cdot 5}(a_0 + a_1x + a_2x^2 + \cdots + a_nx^n + \cdots)^5 \\ &+ \frac{1\cdot 3\cdot 5}{2\cdot 4\cdot 6\cdot 7}(a_0 + a_1x + a_2x^2 + \cdots + a_nx^n + \cdots)^7 + \cdots . \end{aligned} \tag{12}$$

Newton thought that using the **method of undetermined coefficients**[5] in the above expression, he might be able to determine the coefficients a_0, a_1, a_2, $\ldots$. Newton called this approach the **reverse method** of solving the power series.

Some readers may be unfamiliar with the method of undetermined coefficients. Therefore, before we discuss the result of Newton's findings, let us use an example to explain briefly how the method of undetermined coefficients works. Consider the function $y = \tan^{-1} x$, which appeared in Challenge 6. Using the binomial theorem, we have

$$y = \sqrt{1+x} = 1 + \frac{1}{2}x - \frac{1}{8}x^2 + \frac{1}{16}x^3 - \frac{5}{128}x^4 + \cdots . \tag{13}$$

Let us now find the coefficients a_0, a_1, a_2, $\ldots$ by using the method of undetermined coefficients.

First we have

$$y = \sqrt{1+x} \Leftrightarrow x = y^2 - 1.$$

Therefore, if $y = \sqrt{1+x}$ is represented by the power series as

$$y = \sqrt{1+x} = a_0 + a_1x + a_2x^2 + a_3x^3 + \cdots + a_nx^n + \cdots , \tag{14}$$

[5]The method of undetermined coefficients is a tool used to determine the coefficient of any power in a given polynomial by comparing it to the coefficient of the term of the same power in an identical polynomial. For example, if a polynomial $x^3 + 1$ is divided by the polynomial $x^2 - 2x$, then the quotient and the remainder are obtained by setting $x^3 + 1$ to be equal to the identical polynomial $(ax + b) \times (x^2 - 2x) + cx + d$, that is,

$$x^3 + 1 = (ax + b) \times (x^2 - 2x) + cx + d$$

or

$$x^3 + 1 = ax^3 + (b - 2a)x^2 + (c - 2b)x + d.$$

By comparing the coefficient of each term on the left-hand side of this equation with the coefficient at the term of the same power on the right-hand side of the equation, we have

$$a = 1, \quad b - 2a = 0, \quad c - 2b = 0, \quad d = 1.$$

From the above system of equations, we see that the quotient and remainder are given by polynomials $x + 2$ and $4x + 1$, respectively.

then by substituting for y in the equation $x = y^2 - 1$ the sum of the power series in (14), we have

$$x = (a_0 + a_1x + a_2x^2 + a_3x^3 + \cdots + a_nx^n + \cdots)^2 - 1.$$

If we expand the right-hand side of the above expression, we obtain

$$\begin{aligned} x = (a_0^2 - 1) &+ 2a_0a_1x + (a_1^2 + 2a_0a_2)x^2 + 2(a_1a_2 + a_0a_3)x^3 \\ &+ \{a_2^2 + 2(a_0a_4 + a_1a_3)\}x^4 + \cdots . \end{aligned}$$

Since this is an identity with respect to x, comparing the two sides of the equation gives us

$$\begin{aligned} &a_0 - 1 = 0, \\ &2a_0a_1 = 1, \\ &a_1^2 + 2a_0a_2 = 0, \\ &2(a_1a_2 + a_0a_3) = 0, \\ &a_2^2 + 2(a_0a_4 + a_1a_3) = 0, \\ &\qquad\cdots \end{aligned}$$

From the first formula we have $a_0 = 1$. By substituting $a_0 = 1$ in the second formula, we obtain $a_1 = 1/2$. From the third formula we get

$$a_2 = -\frac{a_1^2}{2a_0} = -\frac{1}{2}\left(\frac{1}{2}\right)^2 = -\frac{1}{8}.$$

Similarly we have

$$\begin{aligned} a_3 &= -\frac{a_1a_2}{a_0} = -\frac{1}{2}\left(-\frac{1}{8}\right) = \frac{1}{16}, \\ a_4 &= -\frac{a_2^2 + 2a_1a_3}{2a_0} = -\frac{1}{2}\left\{\left(-\frac{1}{8}\right)^2 + 2\cdot\frac{1}{2}\cdot\frac{1}{16}\right\} = -\frac{5}{128}. \end{aligned}$$

Using this technique of successive substitutions, we obtain

$$a_0 = 1, \quad a_1 = \frac{1}{2}, \quad a_2 = -\frac{1}{8}, \quad a_3 = \frac{1}{16} \quad \text{and} \quad a_4 = -\frac{5}{128},$$

which are identical to the coefficients in equation (13).

If we do not mind spending a lot of time, the method of undetermined coefficients helps us to find finite numbers of coefficients in expression (13) by determining the first coefficient, then the second, and so on. The right-hand side of (13) has infinite number of terms; therefore, unless we find some general rule, we cannot say that $y = \sqrt{1+x}$ is represented by the power series expansion.

One uses the method of undetermined coefficients to obtain the power series by the reverse method. The reverse method is a heuristic method whereby we first find a limited sequence of coefficients, and then make a conjecture about the nth term of the sequence. We then need to prove that our conjecture is valid.

In 1669, Newton solved equation (12) by the method of undetermined coefficients and obtained the first five terms of the power series expansion for $y = \sin x$ as follows:

$$y = x - \frac{1}{6}x^3 + \frac{1}{120}x^5 - \frac{1}{5040}x^7 + \frac{1}{362880}x^9 - \cdots .$$

He then concluded that the coefficient of x^{2n+1} (up to the sign) is $\frac{1}{(2n+1)!}$.

This surprising assertion was valid! The function $y = \sin x$ is indeed represented by the following power series:[6]

$$\sin x = x - \frac{1}{3!} + \frac{1}{5!}x^5 - \frac{1}{7!}x^7 + \frac{1}{9!}x^9 - \cdots + (-1)^n \frac{1}{(2n+1)!}x^{2n+1} + \cdots .$$

This representation is valid for any value of x.

This discovery opened doors to further study in many disparate areas. For example, if we substitute $\sin\frac{\pi}{6} = \frac{1}{2}$, $\sin\frac{\pi}{4} = \frac{1}{\sqrt{2}}$, $\sin\frac{\pi}{2} = 1$, $\sin\pi = 0$ into the above expressions, we obtain from each of the substitutions a mysterious relation expression relating to the circumference ratio π.

Now, let us compare the power series expansions (8) and (10) of the functions $y = \tan^{-1} x$ and $y = \sin^{-1} x$ respectively. We realize that the power series representing $y = \tan^{-1} x$ is much simpler than the power series representing $y = \sin^{-1} x$. Therefore, it seems that we should be able to obtain the power series expansion of the function $y = \tan x$ by solving in reverse the power series expansion of $y = \tan^{-1} x$ just as we obtained the power series expansion of $y = \sin x$ by the method of undetermined coefficients. But in fact, it is not possible to solve the function $y = \tan^{-1} x$ in reverse. We know this because the power series expansion of $y = \tan^{-1} x$ is known today to look as follows:

$$\text{if } |x| < \frac{\pi}{2}, \text{ then } \tan x = \sum_{n=1}^{\infty} \frac{2^{2n}(2^{2n}-1)B_n}{(2n)!}x^{2n-1}.$$

[6] $y = \cos x$ is represented by the power series expansion as follows:

$$\cos x = 1 - \frac{1}{2!}x^2 + \frac{1}{4!}x^4 + \cdots + (-1)^n \frac{1}{(2n)!}x^{2n} + \cdots .$$

The numbers B_n, which appear in the numerator above, are called Bernoulli numbers.[7] The sequence of Bernoulli numbers is an irregular sequence of rational numbers. The first nine terms of the sequence are given as follows:

$$B_1 = \frac{1}{6}, \quad B_2 = \frac{1}{30}, \quad B_3 = \frac{1}{42}, \quad B_4 = \frac{1}{30}, \quad B_5 = \frac{5}{66},$$
$$B_6 = \frac{691}{2730}, \quad B_7 = \frac{7}{6}, \quad B_8 = \frac{3617}{510}, \quad B_9 = \frac{43867}{798}, \quad \ldots .$$

B_{15} is given by the following (rational) monstrosity:

$$B_{15} = \frac{8615841276005}{14322}.$$

The coefficients of terms in the power series expansion of $\tan x$ do not follow any particular rule. Hence, even if we try to employ the reverse method and obtain the values of the first several coefficients, we will not know the coefficient of the general term of the series. Without being able to express the coefficient of the general term of the series, it is impossible to obtain the power series expansion of $\tan x$ from that of $\tan^{-1} x$.

1.6. History

At the beginning of the 17th century, John Wallis (1616–1703) and his contemporaries developed what is known as **infinite operation**. This operation later was expanded by Newton and Leibniz. In the 1680s, these two mathematical geniuses introduced a completely new mathematical system called calculus. Even at that early date, both Newton and Leibniz knew how to represent the functions $\tan^{-1} x$, $\sin^{-1} x$, and $\sin x$ by power series expansions. They also knew the power series representation of the function $y = e^x - 1$. They started from the power series expansion of $\log(1 + x)$, which is given by

$$\begin{aligned}\log(1+x) &= \int_0^x \frac{1}{1+t}\,dt = \int_0^x (1 - t + t^2 - t^3 + t^4 - \cdots)\,dt \\ &= x - \frac{1}{2}x^2 + \frac{1}{3}x^3 - \frac{1}{4}x^4 + \frac{1}{5}x^5 - \cdots .\end{aligned}$$

They then solved the above power series equation in reverse and obtained the following:

$$y = e^x - 1 = x + \frac{1}{2!}x^2 + \frac{1}{3!}x^3 + \frac{1}{4!}x^4 + \cdots .$$

[7]Bernoulli numbers are defined by the following series:

$$\frac{x}{e^x - 1} = 1 - \frac{x}{2} - \sum_{n=1}^{\infty} \frac{(-1)^n B_n x^{2n}}{(2n)!} \quad \text{for } |x| < 2\pi.$$

Since Newton was able to represent a number of functions by the power series, he likely became convinced that any function could, in fact, be represented by a power series. Quite probably, he arrived at this conclusion after observing both the above power series expansions, and also the power series obtained by the general binomial theorem.[8] However, his conjecture was incorrect, as current knowledge about functions reflects. But if we narrow our focus to those functions which appear in Newtonian mechanics, then Newton's conjecture is correct.

In 1715, Brook Taylor (1685–1731) showed that if a function $f(x)$ is represented by a power series expansion, then the power series is given by the higher order derivatives of f. This series is called the **Taylor series** for $f(x)$. In 1742, by taking a special case of the Taylor series expansion, Colin Maclaurin (1698–1746) discovered the following power series expansion for $f(x)$, called the **Maclaurin series**:

$$(*)\ \ f(x) = f(0) + \frac{f'(0)}{1!}x + \frac{f''(0)}{2!}x^2 + \frac{f'''(0)}{3!}x^3 + \cdots + \frac{f^{(n)}(0)}{n!}x^n + \cdots .$$

Most students encounter the above expansion in lecures on differential and integral calculus. However, the general term of this series can only be found once we obtain $f^{(n)}(0)$. Hence, the Maclaurin series for $f(x)$ will be found only when f has derivatives of all orders.

CHALLENGE 7. (For those of you who studied calculus.) Find the Maclaurin series for e^x, $\sin x$, and $\cos x$ using the power series expansion $(*)$.

As we have seen, once we know that a function can be represented by a power series, then the power series is the Taylor series. Representing a function by a power series via integration and differentiation of higher orders, we catch a glimpse of the world of infinite operations. Since the core of that world is the system of differential and integral calculus, we can fairly say that the 18th century was an era of analysis which blossomed from the study of calculus.

Isaac Newton (1642–1727) was born in the remote countryside of eastern England. He was, by all accounts, an ordinary child who spent a great deal of his time tinkering with mechanical devices. He entered the Trinity College of Cambridge University, but in 1665 a plague epidemics spread throughout England. As a result, the University was closed and Newton returned to his home in the English countryside. It was during this forced hiatus that Newton began his deep study of both physics and mathematics. Newton later stated that the time he spent in the countryside while the University was closed was the most creative and productive period of his life. In fact, he discovered universal gravitation, which is amongst his three greatest discoveries ever. During this time, he also established the study of differential and integral calculus and the fundamental idea of the spectral

[8]It is believed that Newton's study of the general binomial theorem was the starting point of his mathematical research.

nature of light. These amazing scientific achievements he incorporated into *Principia*, published in 1687 when Newton was 46 years old.

Newton's research laid the foundation of modern science; his influence is felt at every level of modern technology. Not long ago, Cambridge University created the Isaac Newton Institute for Mathematical Sciences to commemorate his numerous scientific achievements. Today, this institute is world renowned for cutting-edge research in the fields of mathematics and physics.[9]

[9]Toward the end of his life, Newton turned to matters of religion, and he also became the Master of the Royal Mint!

LECTURE 2

Elliptic Functions

If you have not yet studied calculus, you may wish to skip the detailed calculations below and instead concentrate on grasping the main concepts.

2.1. Ellipses

An ellipse is a curve obtained by slicing a cone with a tilted plane as in Figure 2.1. The set of points that satisfy the equation

(1) $$\frac{x^2}{a^2} + \frac{y^2}{b^2} = 1,$$

where $a > b > 0$, is called an ellipse, with $2a$ being the major axis and $2b$ the minor axis.

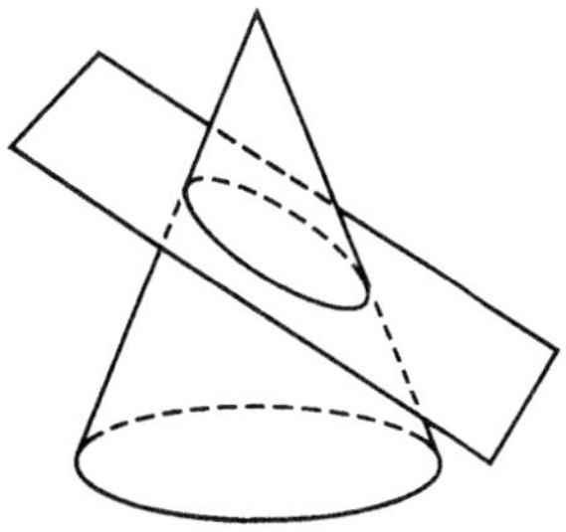

FIGURE 2.1

The equation of an ellipse may be written in the form

$$x^2 + \left(\frac{a}{b}y\right)^2 = a^2.$$

Hence, if a point (x, y) is on the graph of equation (1), then by multiplying the y-coordinate of this point by $\frac{a}{b}$, it will be moved to the circle of radius a centered at the origin.

On the other hand, if a point (x, y) is on the circle of radius a centered at the origin, then its coordinates are expressed as follows:

$$x = a\cos\theta, \quad y = a\sin\theta.$$

If we multiply the y-coordinate of this point by $\frac{b}{a}$, we can express a point (x, y) on the ellipse as

(2) $$x = a\cos\theta, \quad y = b\sin\theta.$$

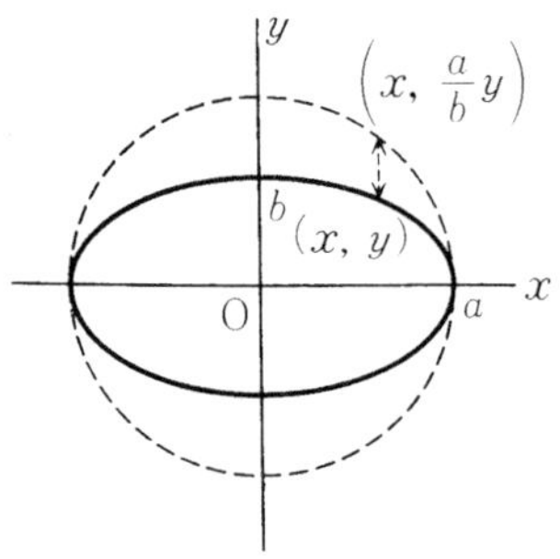

FIGURE 2.2

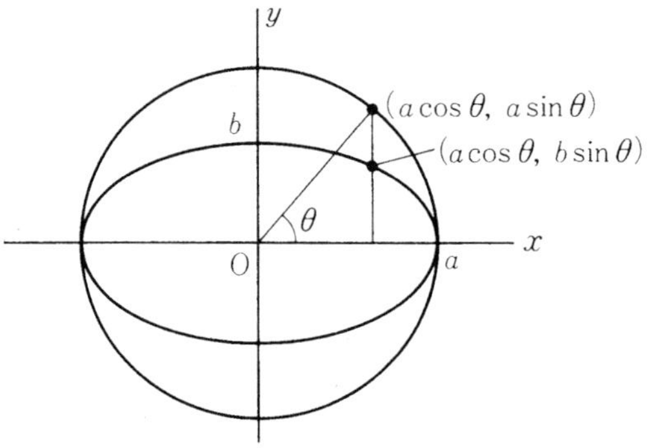

FIGURE 2.3

An ellipse can be described as the set of all points with a given sum of distances from two fixed points. Consider the following two fixed points on the x-axis:

$$F(c,0) \quad \text{and} \quad F'(-c,0)$$

which are symmetric with respect to the origin.

Let P be a point on the plane such that the sum of the distances from P to F and from P to F' is $2a$. That is, the point P satisfies the following relation:

$$(\text{length of } \overline{PF}) + (\text{length of } \overline{PF'}) = 2a.$$

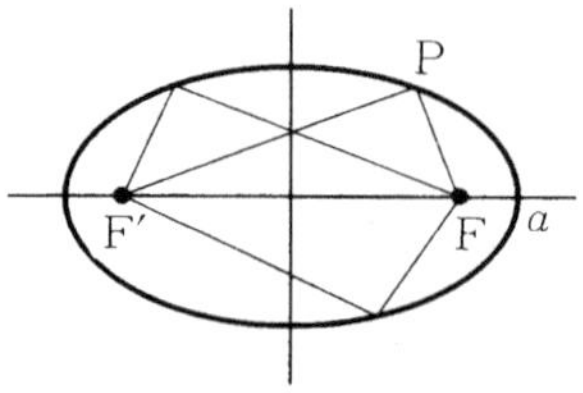

FIGURE 2.4

If the coordinates of P are (x, y), then the above relation is expressed by the following equation:

$$(1') \qquad \sqrt{(x-c)^2+y^2}+\sqrt{(x+c)^2+y^2}=2a.$$

This may be written as

$$\sqrt{(x-c)^2+y^2}=2a-\sqrt{(x+c)^2+y^2}.$$

Squaring both sides of the equation and simplifying the expression, we obtain

$$a\sqrt{(x+c)^2+y^2}=a^2+cx.$$

Squaring both sides of the last equation, we obtain

$$(a^2-c^2)x^2+a^2y^2=a^2(a^2-c^2).$$

Since $a > c > 0$, we can consider the number $b=\sqrt{a^2-c^2}$, so that $b^2 = a^2-c^2$. Then the above equation takes the following form:

$$b^2x^2+a^2y^2=a^2b^2.$$

Dividing both sides by a^2b^2, we obtain equation (1), and the graph of the equation is an ellipse.

CHALLENGE 8. Verify the previous computations.

As we know, the sine function was derived from the study of the circle. We might therefore say that the circle gave birth to a function called sine. This "birth process" gave mathematicians a new point of view, which allowed them to see that a function can be represented by a power series. This new point of view was instrumental in furthering the science of mathematics from the end of the 17th century to the 18th century. What do you suppose would happen if we shift our focus from a circle to an ellipse? Wouldn't it be reasonable to assume that a totally new point of view would also emerge? This is exactly what happened at the end of the 18th and the beginning of the 19th century when mathematicians began to focus their attention on ellipses and curves called lemniscates (we will be discussing lemniscates in the next section). This new focus triggered a major change in the "flow" of the science of mathematics.

When we consider a circle, the question of how to determine the arc length always looms large. Here, we will discuss how to express the arc length of an ellipse by using an integral.

The standard form for the equation of an ellipse is

$$\frac{x^2}{a^2}+\frac{y^2}{b^2}=1 \quad (b>a>0).$$

In our discussion, we will consider an ellipse with the major axis lying on the y-axis, that is, $b > a > 0$ (see Figure 2.5).

If we express this ellipse by using a parameter θ, as we have done earlier for the circle, then we have

$$x=a\cos\theta, \quad y=b\sin\theta.$$

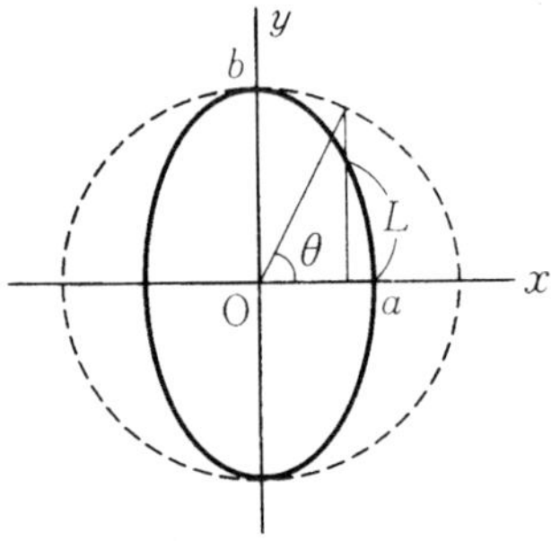

FIGURE 2.5

With these parametric equations, let us obtain the length L of the arc of the ellipse as θ varies from 0 to α (counterclockwise).

♣ In general, the length of the graph of a function $y = f(x)$ from $x = a$ to $x = b$ is given by the definite integral as follows: first divide the interval $[a, b]$ into n subintervals having the same length by choosing numbers $x_0, x_1, x_2, \ldots, x_{n-1}, x_n$ such that, $a_0 = x_0 < x_1 < x_2 < \cdots < x_{n-1} < x_n$; next let Δx denote the length of each subinterval, and let Δy_k denote the difference between the values of f at the end points at each subinterval, that is, $x_{k+1} - x_k = \Delta x$, $f(x_{k+1}) - f(x_k) = \Delta y_k$, $k = 0, 1, 2, \ldots, n-1$. If the graph of $f(x)$ over the interval $[x_k, x_{k+1}]$ is replaced by the segment as illustrated in Figure 2.6, we can approximate the length of the graph of $f(x)$ over this interval by using the Pythagorean theorem, that is,

$$\sqrt{(\Delta x)^2 + (\Delta y_k)^2}. \tag{\#}$$

Therefore, if the graph of $f(x)$ over the interval $[a, b]$ is replaced by a polygonal line where each segment replaces the corresponding section of the graph of $f(x)$, then we can approximate the arc length of $f(x)$ by calculating the length of the polygonal line(s), that is:

$$\sum_{k=0}^{n-1} \sqrt{(\Delta x)^2 + (\Delta y_k)^2} = \sum_{k=0}^{n-1} \sqrt{1 + \left(\frac{\Delta y_k}{\Delta x}\right)^2}\, \Delta x.$$

If n increases without bound, the limit of the sum exists and the arc length of the graph of $f(x)$ is given by

$$\int_a^b \sqrt{1 + \left(\frac{dy}{dx}\right)^2}\, dx.$$

In mathematics it is sometimes said that the infinitesimal length is given by $dx^2 + dy^2$ while referring to expression (#).

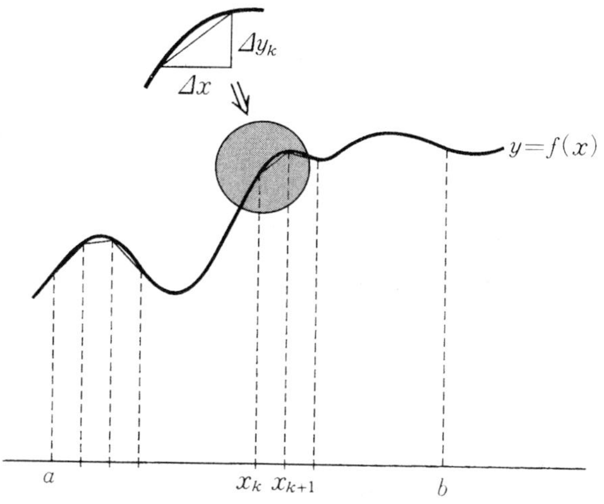

FIGURE 2.6

Of course, when using such a language, the notations dx and dy should be clearly defined in advance. For convenience and for our limited purposes here, we will use these notations without definition.

If we differentiate both sides of the parametric equations in (2) with respect to θ, we obtain

$$dx = -a\sin\theta\, d\theta, \quad dy = b\cos\theta\, d\theta.$$

Therefore,

$$dx^2 + dy^2 = (a^2\sin^2\theta + b^2\cos^2\theta)\, d\theta^2.$$

From this equation, the arc length L of the ellipse, as θ varies from 0 to α, is given by

$$L = \int_0^\alpha \sqrt{a^2\sin^2\theta + b^2\cos^2\theta}\, d\theta = b\int_0^\alpha \sqrt{1 - \left\{1 - \left(\frac{a}{b}\right)^2\right\}\sin^2\theta}\, d\theta.$$

Since we have assumed that $b > a > 0$, we have

$$1 - \left(\frac{a}{b}\right)^2 = k^2$$

for some k.

If we substitute k^2 for $1 - \left(\frac{a}{b}\right)^2$ in the above equation, then, we obtain an integral in the form

$$\int \sqrt{1 - k^2\sin^2\theta}\, d\theta \tag{3}$$

related to the arc length of the ellipse.

To understand this integral, let us substitute t for $\sin\theta$, that is:

$$\sin\theta = t.$$

By differentiating the above equation with respect to t we have

$$\cos\theta d\theta = dt.$$

If the graph of the ellipse is in the first quadrant, we have $\cos\theta > 0$. Hence, we have

$$\sqrt{1-\sin^2\theta}\, d\theta = dt.$$

After making substitutions $t = \sin\theta$ and $d\theta = \frac{dt}{\sqrt{1-\sin^2\theta}}$, we can express the integral (3) as the integral with respect to t,

$$\int \sqrt{\frac{1-k^2t^2}{1-t^2}}\, dt. \tag{4}$$

The integrand in this formula contains the square root and the fraction whose numerator and denominator both are quadratic polynomials. Such an integral can never be expressed in terms of elementary functions (i.e., rational function, radicals, e^x, $\log x$, $\sin x$, $\tan x$, $\sin^{-1} x$, or $\tan^{-1} x$).

♣ The following examples show that if the expression under the radical sign is a quadratic polynomial or its inverse, then the integral can be expressed in terms of elementary functions:

$$\int \frac{1}{\sqrt{1+t^2}}\, dt = \log(t+\sqrt{1+t^2}) + c,$$

$$\int \sqrt{1-t^2}\, dt = \frac{1}{2}(t\sqrt{1-t^2} + \sin^{-1} t) + c,$$

$$\int \sqrt{1+t^2}\, dt = \frac{1}{2}\{t\sqrt{1+t^2} + \log(t+\sqrt{1+t^2})\} + c.$$

On the other hand, the integral (4) can never be expressed in terms of elementary functions.

CHALLENGE 9. Differentiate the right-hand sides of the three equations above, and verify that the resulting derivatives are the functions following the integral sign. The derivative of $\sin^{-1} t$ is $\frac{1}{\sqrt{1-t^2}}$.

This fact caused great excitement amongst 18th century mathematicians (in particular Euler and Lagrange) because up until then no function had been known whose derivative was the integrand in (4). There was no function in any field of the 18th century mathematics such that

$$F'(t) = \sqrt{\frac{1-k^2t^2}{1-t^2}}.$$

Recall that the circle gave rise to trigonometric functions, which led to calculus. In turn, the ellipse has aroused questions in calculus. This line of inquiry has further expanded our understanding of mathematics.

The integral (4) is called an elliptic integral. Note that the integral (4) can be expressed as

$$\int \frac{1}{1-t^2}\sqrt{(1-k^2t^2)(1-t^2)}\, dt.$$

The definition of the general elliptic integral is given by the following.

Let $\varphi(t)$ be either a cubic or fourth degree polynomial in one variable t. If $R(t, \sqrt{\varphi(t)})$ is a rational function of t and $\sqrt{\varphi(t)}$, then the integral

$$\int R(t, \sqrt{\varphi(t)})\, dt$$

is called an elliptic integral.

The study of elliptic integrals is essentially the study of the following three fundamentally different types of integrals:

$$\text{Type I:} \qquad \int \frac{1}{\sqrt{(1-t^2)(1-k^2t^2)}}\, dt.$$

$$\text{Type II:} \qquad \int \sqrt{\frac{1-k^2t^2}{1-t^2}}\, dt.$$

$$\text{Type III:} \qquad \int \frac{1}{(t^2-a^2)\sqrt{(1-t^2)(1-k^2t^2)}}\, dt.$$

The arc length of an ellipse is expressed as an elliptic integral of the second type.

2.2. Lemniscates

As we discussed in the previous section, an ellipse is the set of all points in the plane with the sum of distances to two fixed points equal to a constant value $2a$. Consider now the set of all points in a plane with the product of distances to two fixed points being equal to a constant value K. This set of points describes a curve called a lemniscate. Accordingly, the arc length of a lemniscate is given by an elliptic integral called the lemniscate integral. The study of this integral opened new areas of research for mathematicians of the 19th century.

Let us obtain the equation to describe a lemniscate, and then study the shape of the resulting curve.

Let K (> 0) be a constant. Choose a rectangular coordinate system and consider two fixed points, $P_1(-a, 0)$ and $P_2(a, 0)$. The curve consisting of all points $P(x, y)$ satisfying the equation

$$\sqrt{(x+a)^2+y^2}\sqrt{(x-a)^2+y^2} = K \tag{5}$$

is called a lemniscate. Let us compare equation (5) with equation (1′). We notice that (1′) uses *addition* whereas (5) uses *multiplication*. Squaring equation (5) and rearranging the terms, we obtain the following equation:

$$(x^2+y^2)^2 = 2a^2(x^2-y^2) + K^2 - a^4. \tag{6}$$

Consider the curve that passes through the origin. The condition for the curve to pass through the origin is given by

$$K = a^2$$

(we set $x = y = 0$ in equation (6)). Hence, the equation of a lemniscate passing through the origin is given by

$$(x^2 + y^2)^2 = 2a^2(x^2 - y^2). \tag{7}$$

But this equation does not yet give us a clear picture of the curve. Therefore let us express the above equation in polar coordinates (r, θ), where

$$x = r\cos\theta \quad \text{and} \quad y = r\sin\theta.$$

Then equation (7) becomes the following polar equation:

$$r^2 = 2a^2\cos 2\theta. \tag{8}$$

From this polar equation, we can see that the lemniscate (7) has the form of a horizontal figure eight (see Figure 2.7).

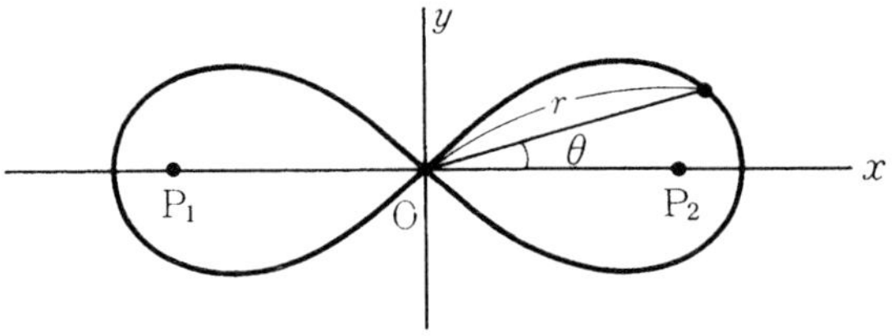

FIGURE 2.7

Let us now compute the length l of the arc between the origin and a point P on the lemniscate in the first quadrant, using the technique called **change of variables**.[1] We have

$$l = \int_0^x \sqrt{1 + \left(\frac{dy}{dx}\right)^2}\, dx = \int_0^r \sqrt{1 + \left(r\frac{d\theta}{dr}\right)^2}\, dr.$$

Next we compute

$$1 + \left(r\frac{d\theta}{dr}\right)^2$$

by the following procedure:

First, differentiate both sides of equation (8) with respect to r to obtain:

$$2r = 2a^2 \cdot 2(-\sin 2\theta)\frac{d\theta}{dr}.$$

[1]Changing variables from x to r for integrals, we have

$$\begin{aligned} dx^2 + dy^2 &= (dr\cos\theta - r\sin\theta\, d\theta)^2 + (dr\sin\theta + r\cos\theta\, d\theta)^2 \\ &= dr^2 + r^2\, d\theta^2. \end{aligned}$$

From this we obtain

$$\int \sqrt{1 + \left(\frac{dy}{dx}\right)^2}\, dx = \int \sqrt{1 + \left(r\frac{d\theta}{dr}\right)^2}\, dr.$$

Therefore,

$$r^2 = 2a^2(-\sin 2\theta) r \frac{d\theta}{dr}.$$

From this equation, we have

$$\left(r\frac{d\theta}{dr}\right)^2 = \frac{r^4}{4a^4 \sin^2 2\theta} = \frac{r^4}{4a^4(1-\cos^2 2\theta)}$$
$$= \frac{r^4}{4a^4 - r^4}$$

(from equation (8)). Therefore,

$$1 + \left(r\frac{d\theta}{dr}\right)^2 = \frac{4a^4}{4a^4 - r^4}.$$

Thus, the arc length l is given by

(9) $$l = \int_0^r \frac{2a^2}{\sqrt{4a^4 - r^4}}\, dr.$$

If we let $a = \frac{1}{\sqrt{2}}$, then equation (7) becomes

$$(x^2 + y^2)^2 = x^2 - y^2,$$

which represents the standard lemniscate. The arc length l of this lemniscate from the origin to the point in the first quadrant whose distance from the origin is s, is given by the integral

(10) $$l = \int_0^s \frac{1}{\sqrt{1 - t^4}}\, dt$$

(the variable r is replaced with t in integral (9)).

Since the integrand contains the square root of a polynomial of degree four, it is an elliptic integral in precisely the same manner as expression (4), which gives the arc length of an ellipse as an elliptic integral. Furthermore, integral (10) is analogous to the following integral:

(11) $$y = \int_0^x \frac{1}{\sqrt{1 - t^2}}\, dt,$$

which represents the arc length $y = \sin^{-1} x$ of a circle (see Figure 2.8). Integral (11) yields the formula $y = \sin^{-1} x$. This provided an impetus to the study of the sine function in greater depth. As you may guess, the study of integral (10) will reveal the subtleties of the relation between (10) and (11).

Let us then consider one aspect of this relation between integrals (10) and (11).

First, in the case of a circle, the arc length in the first quadrant is given by

$$\frac{\pi}{2} = \int_0^1 \frac{1}{\sqrt{1 - t^2}}\, dt.$$

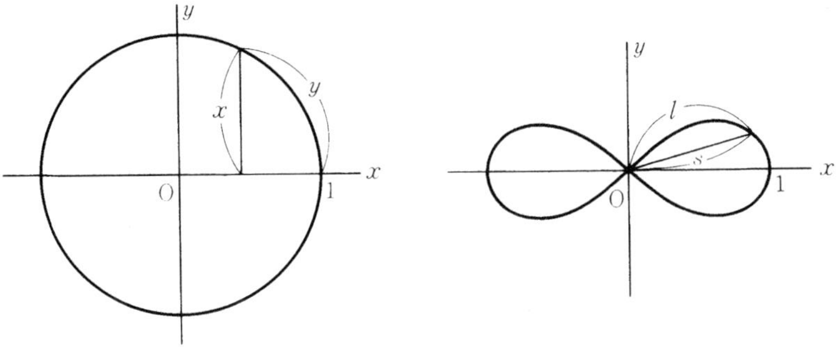

FIGURE 2.8

In the case of an ellipse, the arc length in the first quadrant is given by

$$\omega = \int_0^1 \frac{1}{\sqrt{1-t^4}}\,dt.$$

If we approximate $\frac{\pi}{2}$ and ω to sixteen decimal places, we have

$$\frac{\pi}{2} \approx 1.5707963267948966 \quad \text{and} \quad \omega \approx 1.3110287771460598.$$

As you can see, the arc of a lemniscate is slightly shorter than that of a circle.

In addition, $y = \sin x$ is defined by inverting the following integral:

$$x = \int_0^y \frac{1}{\sqrt{1-t^2}}\,dt$$

with $0 \le y \le 1$, where x and y in equation (11) are interchanged. Analogously, a function called a lemniscate sine, $s = s(x)$, is defined by inverting the following elliptic integral:

$$x = \int_0^s \frac{1}{\sqrt{1-t^4}}\,dt, \tag{12}$$

where $0 \le s \le 1$. Similarly to the formula $\sin \frac{\pi}{2} = 1$, we have $s(\omega) = 1$.

By the binomial theorem,

$$\frac{1}{\sqrt{1-t^4}} = 1 + \frac{1}{2}t^4 + \frac{1\cdot 3}{2\cdot 4}t^8 + \frac{1\cdot 3\cdot 5}{2\cdot 4\cdot 6}t^{12} + \cdots.$$

If we integrate the infinite series as shown on the right-hand side of this expression from 0 to s, then from equation (12) we get

$$x = x(s) = s + \frac{1}{2\cdot 5}s^5 + \frac{1\cdot 3}{2\cdot 4\cdot 9}s^9 + \frac{1\cdot 3\cdot 5}{2\cdot 4\cdot 6\cdot 13}s^{13} + \cdots \quad (|s| < 1).$$

Solving this equation in reverse, we obtain the lemniscate function $s(x)$. The first five terms of the infinite series for $s(x)$ are[2]

$$s(x) = x - \frac{1}{10}x^5 + \frac{1}{120}x^9 - \frac{11}{5600}x^{13} + \frac{217}{353600}x^{17} - \cdots .$$

According to the definition of a lemniscate derived from integral (12), the values of $s(x)$ are determined only for $0 \leq x \leq \omega$. Hence, the correspondence between x and $s(x)$ can be regarded as the correspondence between the arc length of a lemniscate from the origin to a point P and the length of the chord from the origin to the point P as in Figure 2.9 (we assume that $s(x) < 0$ in the third and fourth quadrants). Then we can determine the value of $s(x)$ for $x \geq 0$. For $x < 0$, notice that the above infinite series representation of $s(x)$ contains only odd powers of x, that is, x, x^5, x^{13}, x^{17}, and so on. Therefore, we can define

$$s(-x) = -s(x).$$

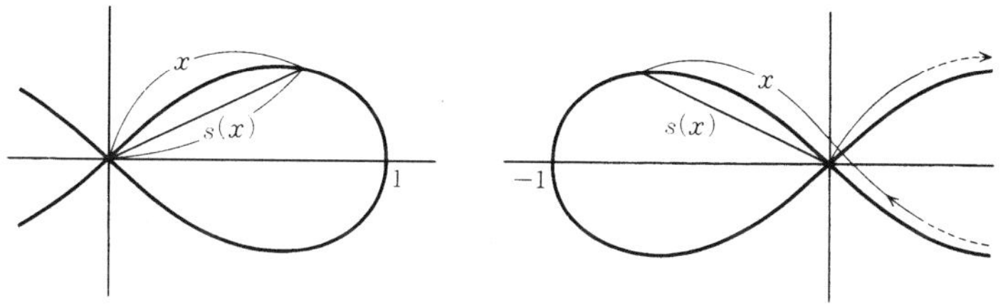

FIGURE 2.9

Since the length of the whole lemniscate is 4ω, if a variable point P on the lemniscate makes a complete turn around the lemniscate, the length of a chord $s(x)$ at the end of the turn will be equal to the original value at the starting point. Thus we have

$$s(x + 4\omega) = s(x).$$

Just as $y = \sin x$ is a periodic function with period 2π, so $y = s(x)$ is a periodic function with period 4ω. Also, as you might have guessed, the graphs of $y = s(x)$ and $y = \sin x$ are waves (Figure 2.10).

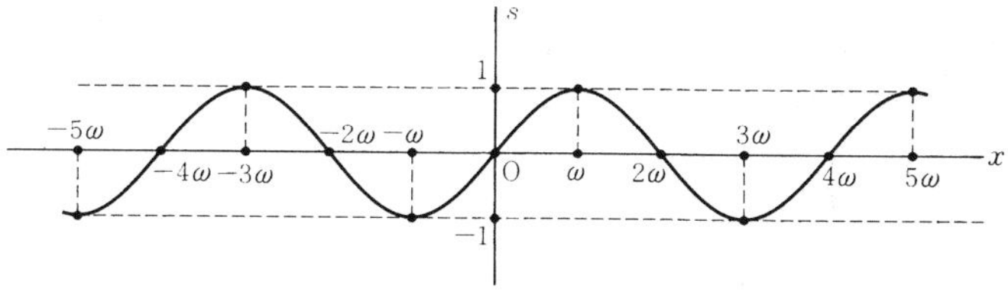

FIGURE 2.10

[2]This infinite series converges if $|x| < \sqrt{2}\omega$, whereas the power series representing $\sin x$ converges for all x. It is differences like this—subtle, but perhaps of great significance—that catch the eye of an eager mathematician.

The lemniscate sine function is a typical elliptic function. In general, a function defined as the inverse function of an elliptic integral of the first type is called an elliptic function.[3]

Let us briefly discuss another aspect of the history of mathematics. Mathematical analysis was developed in the 18th century. By the 19th century, mathematics had reached a fork on the road. One path led to trigonometric functions, and the other path led to elliptic functions. Trigonometry leads to the picture where functions defined on the number line are viewed as compositions of wave functions (using Fourier series). This approach allowed a mathematical understanding of natural phenomena. On the other hand, the study of elliptic functions provided the means to expand the world of mathematical analysis from real numbers to complex numbers. This approach also furthered the integration of the study of number theory and function theory. So, as you can see, the study of trigonometric functions and elliptic functions greatly expanded the scope of the 19th century mathematics.

2.3. Multiple-Angle Formulas

We have been discussing a rather complicated topic, so let us take a short break and discuss some familiar multiple-angle formulas (which you likely know fairly well). We will consider these formulas from a different point of view, thereby enhancing our understanding.

As you know, using the addition formulas for sine and cosine, we obtain the following double angle formulas:

$$\sin 2\theta = 2\sin\theta\cos\theta, \quad \cos 2\theta = 1 - 2\sin^2\theta.$$

Using the first of these double angle formulas we can obtain the triple angle formula for the sine function as follows:

$$\begin{aligned}\sin 3\theta = \sin(2\theta+\theta) &= \sin 2\theta\cos\theta + \sin\theta\cos 2\theta,\\ &= 2\sin\theta\cos^2\theta + \sin\theta - 2\sin^3\theta,\\ &= 2\sin\theta(1-\sin^2\theta) + \sin\theta - 2\sin^3\theta,\\ &= 3\sin\theta - 4\sin^3\theta.\end{aligned}$$

This formula does not look particularly noteworthy. However, as we move forward, it may indeed intrigue you.

[3]The function obtained by inverting the integral

$$x = \int_0^y \frac{1}{\sqrt{(1-t^2)(1-k^2t^2)}}\,dt$$

is called an elliptic function, and is denoted by

$$y = \mathrm{sn}(x).$$

If the parameter k in the previous expression is the imaginary unit i, then the integral yields the lemniscate sine.

Let us look at the three graphs in Figure 2.11. You are undoubtedly familiar with the graph of $y = \sin\theta$. But for some of you, this may be the first time you see the graph of $y = \sin^3\theta$. As you know, we have $0 < \sin\theta < 1$. Therefore, for $0 < \theta < \frac{\pi}{2}$, the cube of $\sin\theta$ is less than $\sin\theta$, that is,

$$\sin^3\theta < \sin\theta.$$

Hence, on the interval $\left(0, \frac{\pi}{2}\right)$, the graph of $y = \sin^3\theta$ lies below the graph of $y = \sin\theta$. When $\theta = \frac{\pi}{2}$, both $y = \sin^3\theta$ and $y = \sin\theta$ attain the value 1. The point of inflection of $y = \sin\theta$ on the interval $\left(0, \frac{\pi}{2}\right)$ is the point at which $\sin\theta = \sqrt{\frac{2}{3}}$. Both graphs have period of 2π, and there is only one swell (the pick of the wave) between 0 and π.

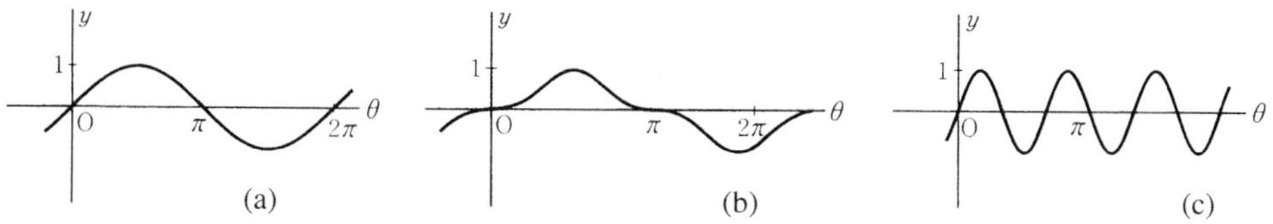

FIGURE 2.11

On the other hand, the graph of $y = \sin 3\theta$ is obtained by three-times contracting of the graph of $y = \sin\theta$. It has two "swells" between 0 and π.

Keeping in mind these facts, let us study the formula for the triple angle. If we triple the amplitude (the height of the wave) of the graph of $\sin\theta$, and substract from the resulting value the amplitude obtained by enlarging the graph of $\sin^3\theta$ four times, we obtain the graph of $y = 3\sin\theta - 4\sin^3\theta$ (that is, $y = \sin 3\theta$). Therefore, the graph of $y = \sin 3\theta$ will have two "swells" between 0 and π, and the shape of the wave will be rather regular. But when we observe the three graphs in Figure 2.11, the fact that $y = 3\sin\theta - 4\sin^3\theta$ has two "swells," does not seem intuitive.

In the past, when mathematicians encountered such nonintuitive problems, they were forced to give up on finding a solution. Now, however, we have tools (such as computer graphing software) that can assist us in solving problems related to drawing graphs of functions. With such software, it is now possible to view the change in the graphs of $y = 3\sin\theta - 4\sin^3\theta$ (as we vary the coefficients).

When we change the coefficients 3 and 4 (even slightly), the orderly periodic wave-like graph disappears. To show what happens for different values of the coefficients, let us introduce the parameter k to the equation $y = 3\sin\theta - 4\sin^3\theta$ as follows:

$$y = 3\sin\theta - k\sin^3\theta,$$

$$y = k\sin\theta - 4\sin^3\theta,$$

$$y = 3k\sin\theta - 4(2-k)\sin^3\theta.$$

The graphs of these functions in three-dimensional space, obtained with the help of a computer, are illustrated in Figure 2.12. Parallel lines on the graphs correspond to the k-axis. If we slice these surfaces at a given value of k, then the curves which appear as cross-sections represent the graphs of equations with this value of k. The waves depicted by white lines are periodic. They represent the graphs of the equation of the triple angle formula.

The three-dimensional picture of graphs of the functions shows different motion of waves in the ocean. It is amazing to think that these different wave shapes are obtained just by slightly changing the parameter k in the formula for $y = \sin 3\theta$. What kind of wave motion can we obtain if we use the quintuple angle formula

$$\sin 5\theta = 5\sin\theta - 20\sin^3\theta + 16\sin^5\theta\,?$$

Following the analogy to the triple angle formula, we vary the coefficients of the above equation. Let us change the coefficients on the right-hand side of the equation to $5k$, $-20k$, and $16k$. The graphs of the resulting functions are shown in Figure 2.13, where k varies from 0.5 to 1.5. When $k = 1$, we obtain the waves depicted by white lines (the graph of the equation of the quintuple angle formula). These three graphs show more complicated motion of waves, similar to the motion of the ocean when a typhoon is approaching.

Next let us study the triple and quintuple angle formulas, $\sin 3\theta$ and $\sin 5\theta$, from a different perspective. We begin with the following equation:

$$\sin 3\theta = 0 \quad (0 \le \theta < 2\pi).$$

This equation means that if the angle θ is tripled, then $\sin 3\theta = 0$, that is, 3θ is either π or 2π (or a multiple thereof). Let us illustrate this situation on the unit circle in Figure 2.14. If $\beta = \frac{\pi}{3}$, then $-\beta = -\frac{\pi}{3}$ $(= \frac{5}{3}\pi)$. Tripling the angles β and $-\beta$, we obtain π and 5π (which becomes π after two revolutions), respectively. If $\alpha = \frac{2\pi}{3}$, then $-\alpha = -\frac{2}{3}\pi$ $(= \frac{4}{3}\pi)$.

Tripling the angles α and $-\alpha$, we obtain 2π and 4π (which becomes 2π after one revolution), respectively. In other words, the points P, P', Q, Q' on the unit circle with angle measures α, $-\alpha$, β, and $-\beta$, respectively, as illustrated in Figure 2.14, are symmetric with respect to the y-axis and they all have the same altitude. The angles β and $-\beta$ become π after tripling, and α and $-\alpha$ become 2π after tripling.

Consider the angles α and $-\alpha$ (2π after tripling). The equation $\sin 3\theta = 0$ gives the value of the sine at the three points with angle measures 0, α, and $-\alpha$ on the unit circle, which divide the circle into three equal arcs.

If we let $t = \sin\theta$ and $\sin 3\theta = 0$, then the triple-angle formula $\sin 3\theta = 3\sin\theta - 4\sin^3\theta$ becomes a polynomial of degree three,

$$3t - 4t^3 = 0.$$

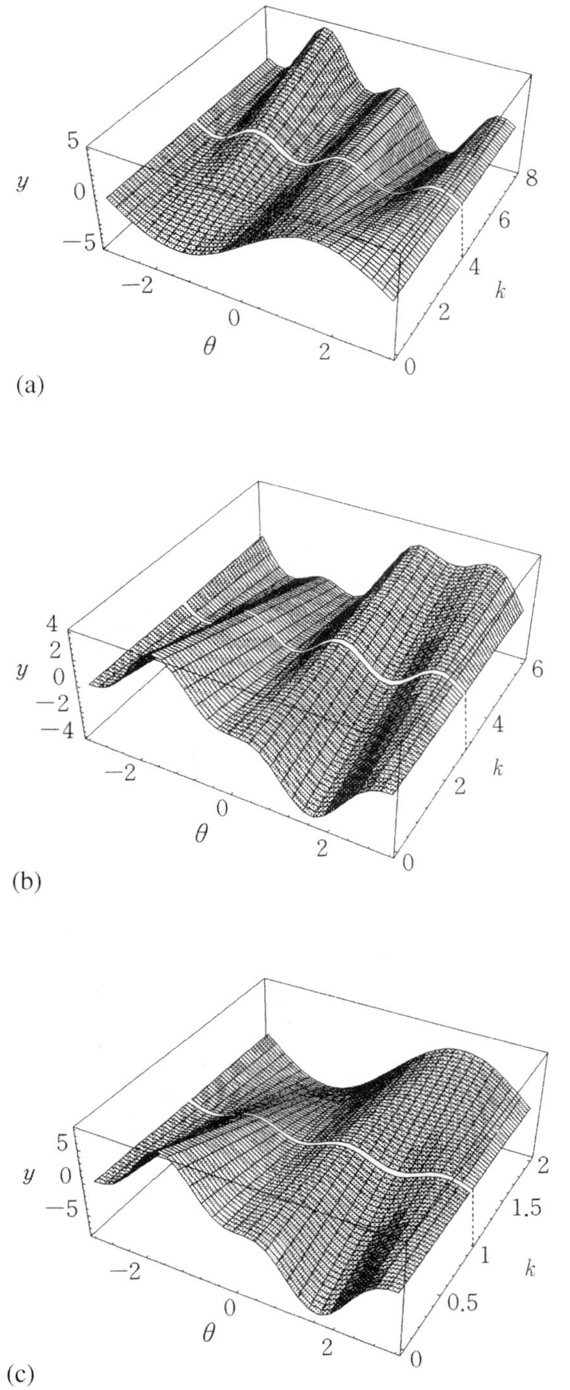

FIGURE 2.12

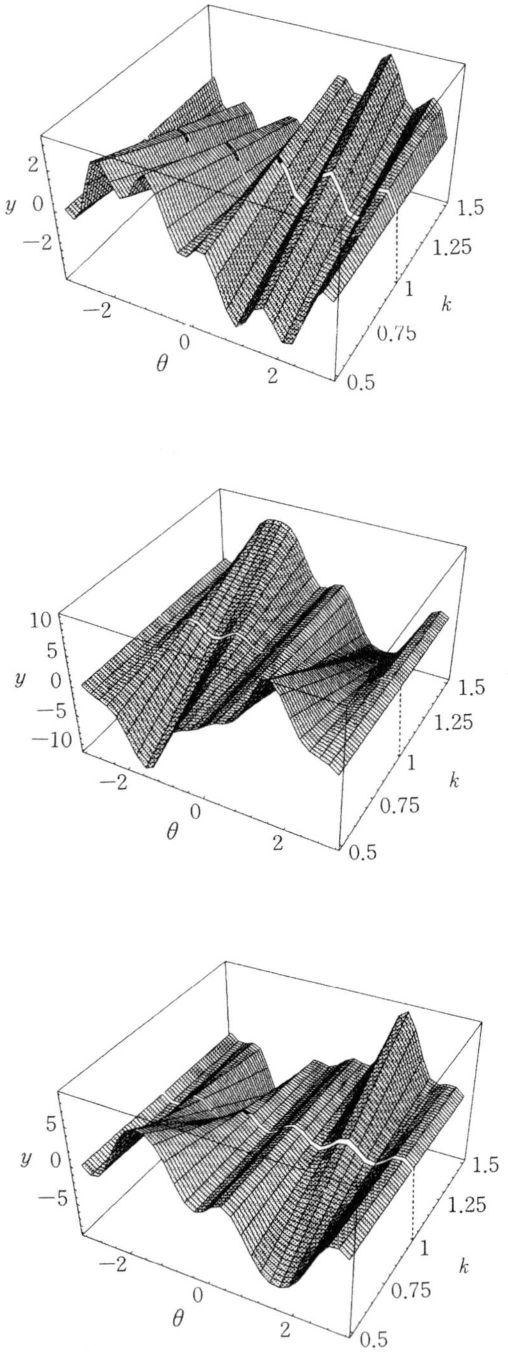

(a)

(b)

(c)

FIGURE 2.13

The solutions of this equation are

$$t = 0,\ \frac{\sqrt{3}}{2},\ -\frac{\sqrt{3}}{2}.$$

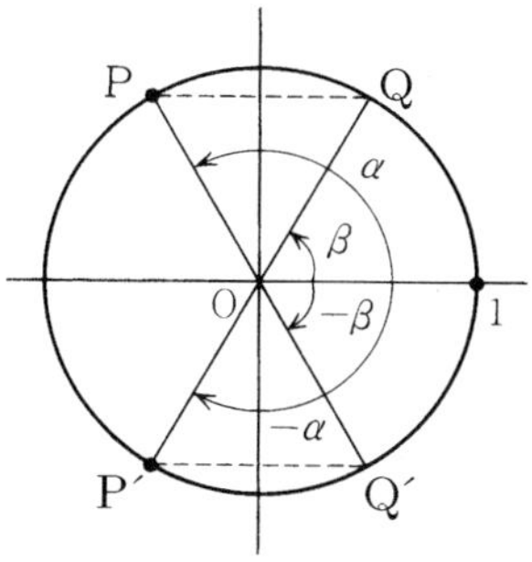

FIGURE 2.14

These three numbers are the values of the sine of 0, $\frac{2\pi}{3}$, and $\frac{4\pi}{3}$, respectively ($\sin 0$, $\sin\frac{2\pi}{3}$, and $\sin\frac{4\pi}{3}$). The coefficients 3 and 4, which appear in the triple-angle formula, relate the cubic equation to the three points on the unit circle. Similarly, if we let $t = \sin\theta$ and $\sin 5\theta = 0$, then the quintuple angle formula

$$\sin 5\theta = 5\sin\theta - 20\sin^3\theta + 16\sin^3\theta$$

becomes the fifth degree polynomial

$$5t - 20t^3 + 16t^5 = 0.$$

The solutions of this equation are the values of the sine at five points on the unit circle which divide the circle into five equal arcs. Let us now verify our previous statement. Since we can write this fifth degree polynomial in the form

$$t(16t^4 - 20t^2 + 5) = 0,$$

we can easily solve the equation and obtain the solutions as

$$t = 0, \quad \pm\sqrt{\frac{10 \pm 2\sqrt{5}}{16}}.$$

If we evaluate the sine function of $\frac{2\pi}{5}$ and $\frac{4\pi}{5}$, we get (see Figure 2.15)

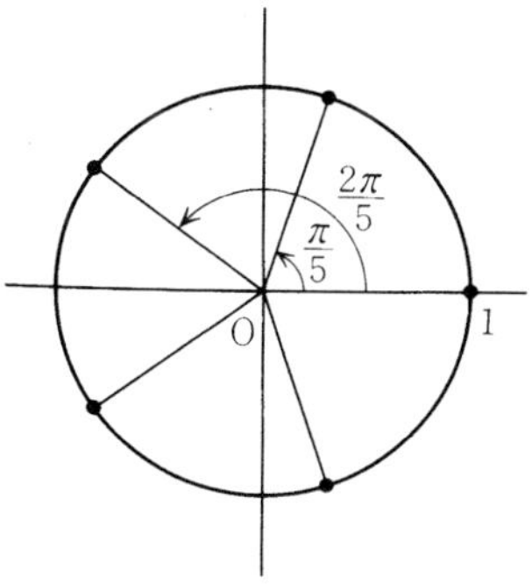

FIGURE 2.15

$$\sin\frac{2\pi}{5} = \sqrt{\frac{10+2\sqrt{5}}{16}} = 0.951056\ldots \quad \text{and}$$

$$\sin\frac{4\pi}{5} = \sqrt{\frac{10-2\sqrt{5}}{16}} = 0.587785\ldots.$$

We may guess that the septimal angle formula gives a seventh degree polynomial in $\sin\theta$, while the solutions of $\sin 7\theta = 0$ give the values of the sine functions evaluated at seven points on the unit circle which divide the circle into seven equal arcs. This is, in fact, correct. We can make similar statements for $\sin 9\theta$, $\sin 11\theta$, and so on.

2.4. The Addition Formula for the Lemniscate Sine

This section should be of interest from a mathematical history perspective.

In Lecture 1, Section 1.2, we have shown how to derive the addition formula for the sine using a geometric approach. There is also the addition formula for the lemniscate sine. However, the proof of the addition formula for the lemniscate sine depends on ideas from analysis, and it is rather difficult to explain in details. Therefore, we will only formulate the result at this time.

First we need to define the lemniscate cosine $c(x)$. The lemniscate cosine is defined as the inverse function to an integral

$$x = \int_c^1 \frac{1}{\sqrt{1-t^4}}\,dt, \tag{13}$$

just as the lemniscate sine is defined as the inverse function of the integral (12). We regard c in equation (13) as a function of x.

The analogy to the cosine in the case of a circle might not appear obvious. However, if $0 \le x \le \frac{\pi}{2}$, we can define $\cos x$ by using the integral

$$x = \int_{\cos x}^1 \frac{1}{\sqrt{1-t^2}}\,dt. \tag{14}$$

The geometric interpretation of this relation is illustrated in Figure 2.16. The illustrations in Figure 2.16 correspond to the illustrations in Figure 1.12 (which show the integral representation of $\sin^{-1} x$).

By studying Figure 2.16, you should be able to explain the relation shown in equation (14).

Further, by observing equation (14), you can see that equation (13) is a natural analogue for the cosine function. In fact, from these observations, we can derive the trigonometric identity

$$\sin^2 x + \cos^2 x = 1$$

and the analogous formula for the lemniscate sine and cosine

$$s(x)^2 + c(x)^2 + c(x)^2 s(x)^2 = 1.$$

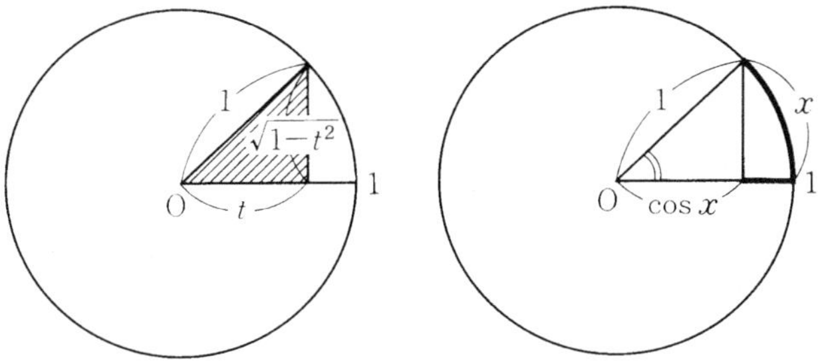

FIGURE 2.16

The addition formula for the lemniscate sine is given by

$$s(x+y) = \frac{s(x)c(y) + c(x)s(y)}{1 - s(x)c(x)s(y)c(y)}.$$

Note that the denominator on the right-hand side of the equation appears to have the form of the addition formula for sine.

Perhaps not surprisingly, Gauss was familiar with the addition formula for the lemniscate sine from his youth. From the addition formula, he derived the following triple and quintuple angle formulas for the lemniscate sine:

$$s(3x) = s\frac{3 - 6s^4 - s^8}{1 + 6s^4 - 3s^8}$$

and

$$s(5x) = s\frac{5 - 2s^4 + s^8}{1 - 2s^4 + 5s^8} \cdot \frac{1 - 12s^4 - 26s^8 + 52s^{12} + s^{16}}{1 + 52s^4 - 26s^8 - 12s^{12} + s^{16}}.$$

The equation $s(3x) = 0$ gives a ninth degree expression, and the equation $s(5x) = 0$ gives a twenty-fifth degree expression. The nineteen year-old Gauss expressed his thoughts about the lemniscate sine function in his diary on March 17, 1797. He asked: "Why is an expression of degree n^2 born out of n equal divisions of a lemniscate?"

The equation $s(3x) = 0$ has three real solutions and six imaginary solutions. The three real solutions correspond to the three equal division points on the lemniscate. This fact is derived in a similar manner as the triple angle formula for a unit circle, in which the equation $\sin 3x = 0$ leads to the three equal division points on the unit circle. But what do the remaining six solutions mean? The equatioin $s(5x) = 0$ has five real solutions, which correspond to the five equal division points on the lemniscate. In addition, there are twenty imaginary solutions. Here, again, we ask, what do these twenty imaginary solutions mean?

From these questions, Gauss made a discovery which led to elliptic functions. Twenty years later, Abel made the same discovery. These questions were answered by extending the notion of the lemniscate sine from the world of real numbers to the world of complex numbers. The answers use the fundamental fact about the nature of elliptic functions, namely that they are doubly periodic on the complex plane.

Carl Friedrich Gauss (1777–1855), considered by many to be the greatest mathematician of the 19th century, made immeasurable contributions to the development of mathematics. However, Gauss was slow to publish his achievements, and when he finally did, he insisted that the finished product be precise and complete in every detail. According to his biography, Gauss was a perfectionist who put forth great effort to insure that each of his discoveries, when revealed, was an equivalent of a masterpiece by a great artist. Any discovery which did not stand up to the standard Gauss had set for his work was not published. He believed that "A cathedral is not a cathedral until the last piece of scaffolding is removed." A seal he used had a picture of a tree which bore only a few fruits, and on it, the following motto was inscribed: "*pauca sed matura*", which translates as "few but ripe." Gauss is remembered as the "prince" of mathematics. Even though he did not publish many of his discoveries (which he felt were not satisfactory), he nevertheless did record them in his diary. His diary entry for January 18, 1797 reads "I have started the study of lemniscates and the integral $\int \frac{1}{\sqrt{1-t^4}}\, dt$."

2.5. Arithmetic and Geometric Means

It $0 < b \leq a$, then the numbers $a_1 = \frac{a+b}{2}$, $b_1 = \sqrt{ab}$ are called the **arithmetic mean** and the **geometric mean** of a and b, respectively. The arithmetic mean of a and b, which is obtained by addition, is half of their sum $a + b$. The geometric mean of a and b, which is obtained by multiplication, is the square root of the product ab. Since a_1 and b_1 represent the average of a and b, a_1 is less than a, and b_1 is greater than b, that is,

$$b \leq b_1, \quad a_1 \leq a.$$

If we subtract b_1 from a_1, we obtain

$$\begin{aligned} a_1 - b_1 &= \frac{1}{2}\,(a + b - 2\sqrt{ab}) \\ &= \frac{1}{2}\,(\sqrt{a} - \sqrt{b})^2 \geq 0. \end{aligned}$$

The inequality $a_1 - b_1 \geq 0$ leads to $b_1 \leq a_1$. Hence we have:

$$\boxed{b \leq b_1 \leq a_1 < a.}$$

In other words, $\sqrt{ab} \leq \frac{a+b}{2}$ for all positive numbers a and b.

If we take a_2 and b_2 as arithmetic and geometric means of a_1 and b_1, respectively, we get

$$a_2 = \frac{a_1 + b_1}{2} \quad \text{and} \quad b_2 = \sqrt{a_1 b_1}$$
$$\text{with} \quad b_1 \leq b_2 \leq a_2 \leq a_1.$$

Repeat this process of taking arithmetic and geometric means as follows. If a_n and b_n are arithmetic and geometric means, then the next arithmetic and geometric means a_{n+1} and b_{n+1} are given by

$$(15) \qquad a_{n+1} = \frac{a_n + b_n}{2} \quad \text{and} \quad b_{n+1} = \sqrt{a_n b_n} \quad (n = 1, 2, 3, \dots).$$

In this way, we obtain an infinite sequence of real numbers a_1, a_2, ..., a_n, ..., b_1, b_2, ..., b_n, ... where

$$b_1 \leq b_2 \leq \cdots \leq b_n \leq \cdots \leq a_n \leq \cdots \leq a_2 \leq a_1.$$

The collection of numbers a_1, a_2, ..., a_n, ... is a monotone nonincreasing sequence with all terms greater than or equal to b_1. In other words, it is a bounded sequence. The collection of numbers b_1, b_2, ..., b_n, ..., is a monotone nondecreasing sequence with all terms less than or equal to a_1. This means that this sequence is also bounded. Since each of these sequences is a bounded monotone infinite sequence, we know that each has a limit α and β, respectively, written as

$$(16) \qquad \lim_{n\to\infty} a_n = \alpha \quad \text{and} \quad \lim_{n\to\infty} b_n = \beta.$$

Therefore, if we increase n ($n \to \infty$) in equation (15), we obtain

$$\alpha = \frac{\alpha + \beta}{2} \quad \text{and} \quad \beta = \sqrt{\alpha\beta}.$$

Hence, from the above two equations, we have

$$\alpha = \beta.$$

Starting from two positive numbers a and b, we find the first arithmetic and geometric means, then the second, and so on. The sequences of successive arithmetic and geometric means have the same limit. This common limiting value is called the arithmetic-geometric mean of a and b and is denoted by $M(a, b)$ (Figure 2.17).

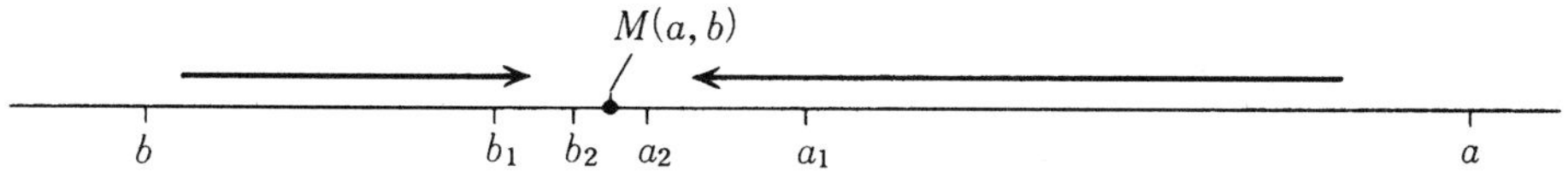

FIGURE 2.17

Note that the convergence rate (the rate of the decrease of a_n and the rate of the increase of b_n) in expressions in (16) is extremely fast. For example, let $a = 100$ and $b = 10$ and compute $a_1, b_1; a_2, b_2; \ldots$. We have

$$\begin{aligned} a_1 &= 55, & b_1 &= 10.4880884\ldots, \\ a_2 &= 60.2440442\ldots, & b_2 &= 24.0175949\ldots, \\ a_3 &= 42.1308195\ldots, & b_3 &= 38.0383628\ldots, \\ a_4 &= 40.0845912\ldots, & b_4 &= 40.0323294\ldots, \\ a_5 &= 40.0584603\ldots, & b_5 &= 40.0584518\ldots, \\ a_6 &= 40.0584561\ldots, & b_6 &= 40.0584561\ldots. \end{aligned}$$

After six repeated operations, a_6 and b_6 are approximately equal (denoted by $a_6 \approx b_6$) and the approximation is accurate up to seven decimal places. This shows that

$$M(100, 10) = 40.0584561\ldots.$$

We can examine how quickly the sequence converges:

$$\begin{aligned} a_{n+1} - b_{n+1} &= \frac{a_n + b_n}{2} - \sqrt{a_n b_n} \\ &= \frac{1}{2}(\sqrt{a_n} - \sqrt{b_n})^2 = \frac{1}{2}\frac{(a_n - b_n)^2}{(\sqrt{a_n} + \sqrt{b_n})^2} \leq \frac{1}{8b}(a_n - b_n)^2. \end{aligned}$$

For example, if $a_n \approx b_n$ is accurate up to four decimal places, then the error $a_n - b_n$ is approximately $\frac{1}{10^5}$. In the next step, the above inequality tells us the error $a_{n+1} - b_{n+1}$ is less than $\left(\frac{1}{10^5}\right)^2 = \frac{1}{10^{25}}$. This means the approximation $a_{n+1} \approx b_{n+1}$ is accurate up to twenty-four decimal places.

The arithmetic-geometric mean has the following properties:

(a) If $\lambda > 0$, then $\lambda M(a, b) = M(\lambda a, \lambda b)$.
(b) $M(a, b) = M\left(\frac{a+b}{2}, \sqrt{ab}\right)$.

The proof of part (a) is obtained by setting $\lambda\left(\frac{a_n+b_n}{2}\right) = \frac{\lambda a_n + \lambda b_n}{2}$ or $\lambda\sqrt{a_n b_n} = \sqrt{\lambda a_n \lambda b_n}$, then letting n increase $(n \to \infty)$. The equation in part (b) is obtained simply by observing the fact that the sequence of operations to obtain the arithmetic-geometric mean of a and b can be started from a and b or $a_1 = \frac{a+b}{2}$ and $b_1 = \sqrt{ab}$. From the properties (a) and (b) we have the following equations:

(c) $M(1, b) = M\left(\frac{1+b}{2}, \sqrt{b}\right) = \frac{1+b}{2} M\left(1, \frac{2\sqrt{b}}{1+b}\right)$.

On May 30, 1799, at the age of twenty-two, Gauss recorded in his journal that he had verified that the approximation

$$\frac{1}{M(1, \sqrt{2})} \approx \frac{2}{\pi}\int_0^1 \frac{1}{\sqrt{1-t^4}}\,dt = \frac{2\omega}{\pi}$$

is accurate up to at least eleven decimal places. Gauss conjectured that

$$M(1, \sqrt{2}) = \frac{\pi}{2\omega} \tag{17}$$

and he believed that if, in fact, this conjecture were proven, then a new field in mathematical analysis would develop. With this belief he entered into the study of elliptic functions and what is now called modular functions.

Most mathematicians, when they first encounter equation (17), seem to sense something mysterious about it. Indeed, $M(1, \sqrt{2})$ gives the arithmetic-geometric mean of the side length of a unit square and the length of a diagonal of that square. The numerical value of $M(1, \sqrt{2})$ is approximately 1.19814023. If the Gauss conjecture were correct, then this numerical value would be equal to

$$\frac{\pi}{\omega} = \frac{2\pi}{4\omega} = \frac{\text{arc length of a unit circle}}{\text{arc length of a lemniscate}}.$$

♣ If $a = \sqrt{2}$ and $b = 1$, then the arithmetic-geometric mean of a and b and the numerical value of $\frac{\pi}{2\omega}$ are approximated to thirteen decimal places as follows:

$$\frac{\pi}{2\omega} = 1.1981402347355\ldots$$

and

$$\begin{aligned} a &= 1.4142135623731\ldots, & b &= 1, \\ a_1 &= 1.2071067811865\ldots, & b_1 &= 1.1892071150027\ldots, \\ a_2 &= 1.1981569480946\ldots, & b_2 &= 1.1981235214931\ldots, \\ a_3 &= 1.1981402347938\ldots, & b_3 &= 1.1981402346773\ldots, \\ a_4 &= 1.1981402347355\ldots, & b_4 &= 1.1981402347355\ldots. \end{aligned}$$

After taking arithmetic and geometric means four times, the approximation $a_n \approx b_n$ is accurate up to at least thirteen decimal places.

Although Gauss never published his work related to the arithmetic-geometric mean, you can find the following theorem in the third volume of his Collected Works.

THEOREM.

$$\frac{1}{M(1,x)} = \frac{2}{\pi} \int_0^{\frac{\pi}{2}} \frac{d\theta}{\sqrt{1-(1-x^2)\sin^2\theta}}.$$

The integral on the right-hand side of the above equation appears slightly different from the one Gauss mentioned in his journal. However, the two integrals are, in fact, equal. To demonstrate this, let us first rewrite the integral that appeared in Gauss' journal:

$$\int_0^1 \frac{1}{\sqrt{1-t^4}}\, dt = \int_0^1 \frac{1}{\sqrt{(1-t^2)(1+t^2)}}\, dt.$$

If we let $t = \sin\theta$, then

$$dt = \cos\theta\, d\theta = \sqrt{1-t^2}\, d\theta.$$

After making the substitutions $t = \sin\theta$ and $dt = \sqrt{1-t^2}\, d\theta$, we obtain

$$\int_0^1 \frac{1}{\sqrt{1-t^2}}\, dt = \int_0^{\frac{\pi}{2}} \frac{1}{\sqrt{1+\sin^2\theta}}\, d\theta.$$

Next, if we let $x = \sqrt{2}$ in the integral in the theorem shown above, we obtain

$$\int_0^{\frac{\pi}{2}} \frac{d\theta}{\sqrt{1+\sin^2\theta}}.$$

This integral is equal to the integral in Gauss' journal.

Gauss' proof of the above theorem involves a complicated set of computations. Therefore we will introduce a slightly different theorem which allows for a simpler proof. Be aware, however, that though this proof will be simpler than that used by Gauss, it still requires a significant number of computations. If you have not yet mastered the method of substitution, simply follow the flow of the calculations.

THEOREM.

$$\frac{1}{M(a,b)} = \frac{2}{\pi} \int_0^{\frac{\pi}{2}} \frac{1}{\sqrt{a^2\cos^2\theta + b^2\sin^2\theta}}\, d\theta.$$

(Note that the expression under the radical sign on the right-hand side of this equation can be written as $a^2 - (a^2 - b^2)\sin^2\theta$. If we substitute 1 for a, and x for b, then the resulting integral is the same as the integral mentioned in Gauss' theorem.)

PROOF. If we let $T(a,b) = \frac{2}{\pi}\int_0^{\frac{2}{\pi}} \frac{1}{\sqrt{a^2\cos^2\theta+b^2\sin^2\theta}}\, d\theta$, then the following equation is the one we need to prove:

$$\frac{1}{M(a,b)} = T(a,b).$$

First, notice that for any positive number α, we have

$$T(\alpha,\alpha) = \frac{1}{\alpha}. \tag{18}$$

Indeed, for $a > 0$ we have

$$\begin{aligned} T(\alpha,\alpha) &= \frac{2}{\pi}\int_0^{\frac{\pi}{2}} \frac{1}{\sqrt{\alpha^2\cos^2\theta + \alpha^2\sin^2\theta}}\, d\theta \\ &= \frac{2}{\pi}\frac{1}{\alpha}\int_0^{\frac{\pi}{2}} \frac{1}{\sqrt{\cos^2\theta+\sin^2\theta}}\, d\theta \\ &= \frac{2}{\pi}\frac{1}{\alpha}\frac{\pi}{2} \\ &= \frac{1}{\alpha}. \end{aligned}$$

If we let $t = b\tan\theta$ and change the variable in the above integral from θ to t, then we obtain

$$\begin{aligned} a^2\cos^2\theta + b^2\sin^2\theta &= \left(a^2 + b^2\frac{\sin^2\theta}{\cos^2\theta}\right)\cos^2\theta \\ &= (a^2+t^2)\cos^2\theta \end{aligned}$$

and

$$dt = b\frac{1}{\cos^2\theta\, d\theta}, \quad \text{that is,} \quad d\theta = \frac{1}{b}\cos^2\theta\, dt.$$

After making the above substitutions, we obtain

$$\begin{aligned} T(a,b) &= \frac{2}{\pi}\int_0^\infty \frac{1}{\sqrt{(a^2+t^2)\cos^2\theta}}\frac{1}{b}\cos^2\theta\, dt \\ &= \frac{2}{\pi}\int_0^\infty \frac{1}{\sqrt{a^2+t^2}}\cdot\frac{1}{b}\cos\theta\, dt \\ &= \frac{2}{\pi}\int_0^\infty \frac{1}{\sqrt{a^2+t^2}b\sqrt{1-\tan^2\theta}}dt \qquad (19) \\ &= \frac{2}{\pi}\int_0^\infty \frac{1}{\sqrt{a^2+t^2}\sqrt{b^2+t^2}}dt \\ &= \frac{1}{\pi}\int_{-\infty}^\infty \frac{1}{\sqrt{a^2+t^2}\sqrt{b^2+t^2}}dt. \end{aligned}$$

(The last equality holds because the integrand is an even function.)

Furthermore, if we let

$$u = \frac{1}{2}\left(t - \frac{ab}{t}\right),$$

then when t increases from 0 to $+\infty$, u increases from $-\infty$ to $+\infty$ monotonically (Figure 2.18).

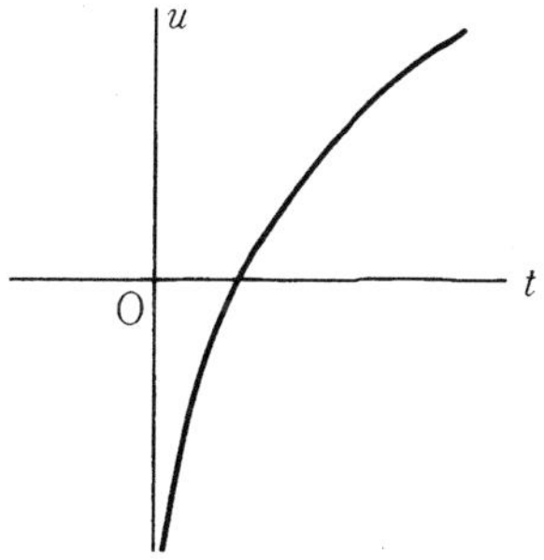

FIGURE 2.18

By letting

$$a_1 = \frac{a+b}{2} \quad \text{and} \quad b_1 = \sqrt{ab},$$

we have

$$\begin{aligned}u^2 + a_1^2 = u^2 + \left(\frac{a+b}{2}\right)^2 &= \frac{(t^2 - ab)^2 + (a+b)^2 t^2}{4t^2} \\ &= \frac{(t^2 + a^2)(t^2 + b^2)}{4t^2}, \\ u^2 + b_1^2 = u^2 + ab &= \frac{(t^2 - ab)^2 + 4abt^2}{4t^2} = \frac{(t^2 + ab)^2}{4t^2}\end{aligned}$$

and

$$du = \frac{t^2 + ab}{2t^2}\, dt.$$

If we substitute u for t in expression (19), and let $a = a_1$, $b = b_1$, we obtain

$$\begin{aligned}T(a_1, b_1) &= \frac{1}{\pi} \int_{-\infty}^{\infty} \frac{1}{\sqrt{a_1^2 + u^2}\sqrt{b_1^2 + u^2}}\, du \\ &= \frac{1}{\pi} \int_{0}^{\infty} \frac{\frac{1}{2t^2}}{\sqrt{t^2 + a^2}\sqrt{t^2 + b^2}\,\frac{1}{4t^2}}\, dt \\ &= \frac{2}{\pi} \int_{0}^{\infty} \frac{1}{\sqrt{t^2 + a^2}\sqrt{t^2 + b^2}}\, dt \\ &= \frac{1}{\pi} \int_{-\infty}^{\infty} \frac{1}{\sqrt{t^2 + a^2}\sqrt{t^2 + b^2}}\, dt \\ &= T(a, b).\end{aligned}$$

Therefore, we have the following sequence of equations:

$$T(a, b) = T(a_1, b_1) = T(a_2, b_2) = \cdots = T(M(a, b), M(a, b)).$$

By substituting $M(a, b)$ for α in expression (18), we obtain

$$T(M(a, b), M(a, b)) = \frac{1}{M(a, b)}.$$

Hence, we have shown the relation

$$T(a, b) = \frac{1}{M(a, b)}$$

and we have proven the theorem. □

Although the theorem has been proven, you may well yet wonder how and why is $M(1, \sqrt{2})$ related to the arc lengths of a unit circle and a lemniscate? We share this puzzlement, and indeed, even the proof written by Gauss does not shed light on this question!

The content of the Gauss theorem likely has a much deeper meaning than what appears on the surface. We have yet to grasp all its subtleties. No one knows whether we can attain Gauss' depth of understanding of this theorem. Sometimes in mathematics we encounter situations in which the truth is presented in many layers. The relation between the arithmetic-geometric mean and a unit circle and a lemniscate, might be one of those situations where the truth is deeper than our ability to understand it.

Chapter 4

Intersection of Geometry and Algebra

Introduction

What enters your mind when your hear the word "geometry"? Among the many things you have learned in high school (or perhaps in junior high) mathematics classes were the properties and characteristics of congruent triangles, circles, and parallel lines. Starting with basic properties of geometric figures, you gradually expanded your understanding of these figures by exploring more complicated properties. Geometry itself has developed in much the same gradual fashion, over many centuries. It is as though we are climbing a mountain of knowledge by carving out the footing for each step with the knowledge gained at the previous ones. Some of the properties at higher points on the "mountain" are so complicated and mysterious, that they appear almost impossible to understand.

Euclid provided the impetus for the development of geometry as a mathematical discipline some 2300 years ago. His work, which is called *Elements*, consists of thirteen volumes, and they offer rigorous proofs of various properties of geometric figures such as triangles and circles. These proofs were based on a set of axioms. Geometry developed by Euclid is called, not surprisingly, Euclidean geometry, and it was considered a model of mathematics in Europe throughout the Middle Ages. Another important step in the evolution of geometry began with the development in algebra that introduced the concept of a coordinate system on a given plane, and the representation of geometric figures by algebraic equations. René Descartes (1596–1650) introduced the notion of a coordinate system in the beginning of the 17th century. For example, a line with slope 2 is expressed as a function of the form $y = 2x + 5$. However, Descartes preferred to use this equation in the form $2x - y + 5 = 0$. The equation of a circle of radius 1 centered at the origin is $x^2 + y^2 - 1 = 0$. As some of you might recall, the condition for a line and a circle to have one point of intersection (the problem of finding a line tangent to a circle) is obtained from the condition that the discriminant of the quadratic equation is equal to zero. In analytic geometry, relations between geometric figures are expressed as relations between equations.

The relations just discussed referred specifically to circles. However, using equations we can also shed light on the properties of ellipses as geometric figures. Among the many properties of ellipses, the following example illustrates one of the most significant properties. Imagine there are two ellipses, one inside the other. Furthermore, imagine there is a triangle inscribed in the outer ellipse and circumscribed about the inner ellipse (Figure 0.1). If we attempt to construct this figure with two ellipses, we will find it to be a difficult task. To see this, try to draw the figure we described by using ellipses. However, by substituting circles for the ellipses in the previous example, it becomes a simple matter of drawing a triangle and its circumscribed and inscribed circles. In the case of ellipses, there is no standard method to construct a triangle inscribed in an ellipse and circumscribed about another ellipse. However, having said that, suppose that you were able, through random trial and error, to draw a triangle inscribed in a given ellipse and circumscribed about the other given ellipse. Given this, let us analyze the following procedure.

Choose a point P on the outer ellipse and then draw a triangle which is circumscribed about the inner ellipse with P as one of its vertices. Surprisingly, the triangle you have just drawn is completely inside the outer ellipse, and in fact, it is inscribed in it.

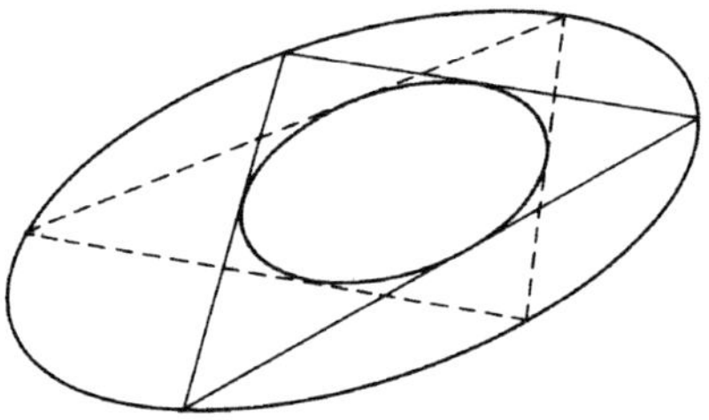

FIGURE 0.1

Now, let us take an example from real life. Start by laying out two ellipses in an open area, one inside the other. Next, let us choose a point on the outer ellipse as the starting point. From this point, walk along the path which is a straight line touching the inner ellipse and proceed along it until you come to a point where the path intersects the outer ellipse. Stop at this point. From this point, choose another straight path tangent to the inner ellipse and proceed along it. You come to another point of intersection on the outer ellipse. Stop again. Change directions the third time and walk along yet another straight line tangent to the inner ellipse until you reach a point on the outer ellipse. If you find that you returned to the original starting point, thus closing the triangle formed by your paths, then, no matter where on the outer ellipse you start, you will always end up at the starting point (after walking in the three distinct directions following the procedure described above).

Do you know which of the properties of ellipses is responsible for this curious phenomenon? Actually, the above phenomenon describes one of the settings of the Poncelet Closure Theorem, which we will discuss in this chapter. The foundation necessary to discuss the Poncelet theorem is that equations for ellipses are given by quadratic polynomials, and the knowledge of basic properties of quadratic equations. As we discuss various proofs of the Poncelet theorem, we think you will come to the realization that observing geometric figures from the algebraic perspective is a powerful tool for clarifying properties of geometric figures.

However, in the algebraic proof of the Poncelet theorem, we will discuss only the case where quadratic equations have real solutions. As you are aware, if the discriminant of a quadratic equation is negative, it has complex solutions. When we attempt to explore corresponding situations in terms of geometric figures, we will see that complex numbers permeate through geometry.

After observing the relationship between geometric figures and equations, many prominent 19th century mathematicians began to see the need to introduce complex numbers into geometry in order to deepen the understanding of the properties of geometric figures. From that new point of view, they could see the relations between variables in the quadratic equations from a geometric standpoint. This new point of view was carried forward into the 20th century mathematics, and ultimately became one of the paramount principles of modern mathematics, called Algebraic Geometry.

Therefore, what we discuss in this chapter is, in a sense, an introduction to algebraic geometry. The topic comes from an era-long past, and may thus seem nothing more than a cast-off shell. But, just as when we turn over a shell and the sun illuminates the pearl-colored beauty of its inner surface, so does the bright light of modern mathematics illuminate our discoveries from mathematics of the 19th century.

In the case where circles are used in place of ellipses, we can prove the Poncelet theorem using the periodicy of elliptic integrals. You will find this proof in Lecture 2.

We also can prove the Poncelet theorem for circles within the framework of plane geometry using properties of coaxial circles. This geometric proof is very interesting, but since this method of proof is somewhat outside of the scope of our lectures, we did not include the proof in our lectures at Kyoto University. However, since the Poncelet theorem is such an elegant theorem in plane geometry, we decided to include this geometric proof in this book as an appendix. Each proof given in this book has its own flavor, showing us the depth of the Poncelet theorem.

LECTURE 1

The Poncelet Closure Theorem

1.1. Poncelet Theorem—the Basics

The main theme of this lecture is the Poncelet theorem. It was Euler who first discovered a theorem from which the Poncelet theorem ultimately was derived. (To give credit where credit is due, this theorem was independently proven by Chapple in 1746.) Euler used plane geometry to prove the following theorem.

EULER'S THEOREM. *Let $\widetilde{C}$ be the circumscribed circle about a triangle $\triangle ABC$ and $\widetilde{C}'$ the inscribed circle in the triangle $\triangle ABC$. Let P be an arbitrary point on the outer circle $\widetilde{C}$. Then there exists a triangle with P as one of its vertices such that $\widetilde{C}$ is circumscribed about it and $\widetilde{C}'$ is inscribed in it.*

If R and r are the radii of $\widetilde{C}$ and $\widetilde{C}'$, respectively, and d is the distance between the centers of $\widetilde{C}$ and $\widetilde{C}'$, then

$$R^2 - d^2 = 2Rr.$$

The first part of Euler's theorem can be reformulated as follows:

Given two circles $\widetilde{C}$ and $\widetilde{C}'$, where $\widetilde{C}'$ is inside $\widetilde{C}$, pick a point A on the outer circle $\widetilde{C}$, and from A draw a line tangent to the inner circle $\widetilde{C}'$. Let B be the point on $\widetilde{C}$ where the tangent line and $\widetilde{C}$ intersect. From B, draw another line tangent to the circle $\widetilde{C}'$ and let C be the point on the outer circle $\widetilde{C}$ where the second tangent line to $\widetilde{C}'$ intersects $\widetilde{C}$. Similarly, draw a third line tangent to the inner circle $\widetilde{C}'$. Suppose the third tangent line intersects the outer circle $\widetilde{C}$ at the starting point A. Then, if we apply the above operation of drawing three tangent lines to the inner circle $\widetilde{C}'$, the third tangent will always return back to the starting point regardless of which point on $\widetilde{C}$ was chosen as the starting point. (See Figure 1.1.)

To restate the above in a "real life" situation, imagine a race track bordered by an inner and an outer circular fence. Suppose a rider on a horse starts moving from a point near the outer fence towards the inner fence along a line which is tangent to the inner fence. When the horse reaches the outer fence again, the rider turns the horse to move along another line which is tangent to the inner fence. If, upon repeating this process for the third time, the horse and rider return to their original starting point, thus closing the triangle formed by their paths, then, as long as the circles remain in the same position relative to one another, no matter where on the outer

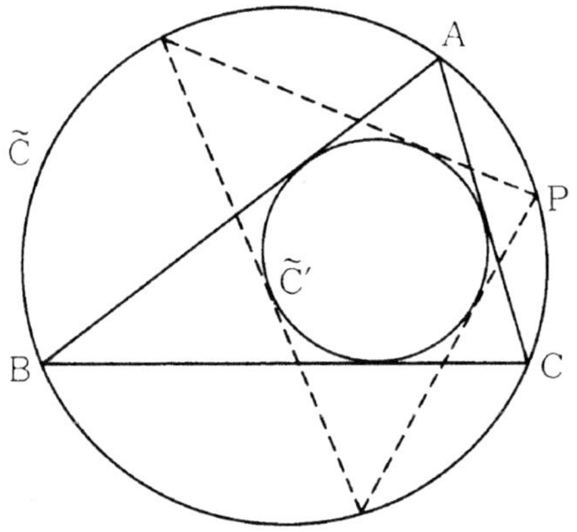

FIGURE 1.1

circle the horse and rider start, they will always end up at the starting point. (Note that the resulting triangular figure obtained by tracing the path of the horse and rider is *closed*. It is this phenomenon of "closing" that Poncelet extracted from Euler's theorem to come out with his own more general theorem.)

Assuming that the first part of the theorem is valid, consider the following problems.

CHALLENGE 1. (i) Verify that to prove the equation $R^2 - d^2 = 2Rr$ it is sufficient to consider only the case where the triangle $\triangle ABC$ is isosceles.

(ii) If $\angle ABC = \theta$, show that

$$(R + d)\cos\theta = r.$$

Hint: Look at the triangle $\triangle AEO'$ in Figure 1.2.

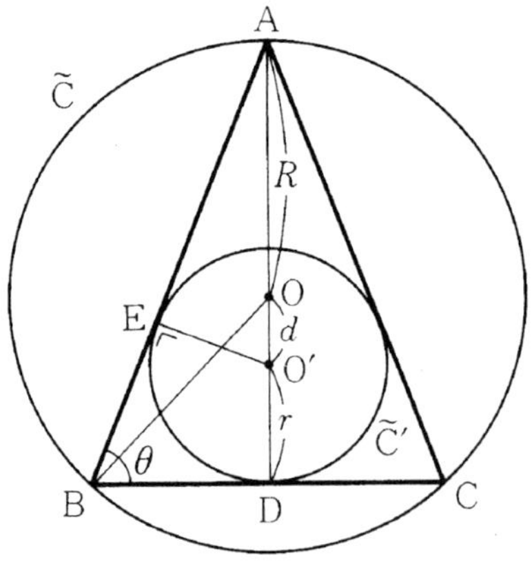

FIGURE 1.2

(iii) Show that $\angle OBD = 2\theta - 90°$.

(iv) Show that $R\sin(2\theta - 90°) = d + r$.

(v) Change the sin notation to cos in the equation in part (iv). By using the double angle formula show that

$$R(1 - 2\cos^2\theta) = d + r.$$

(vi) Express $\cos\theta$ in terms of R, r, and d by using the equation in part (ii). Now substitute the resulting expression for $\cos\theta$ in the equation in part (v) and derive the relation between R, r, and d.

(vii) By rearranging the terms of the above relation show that

$$(R^2 - d^2)(R + d + r) = 2Rr(R + d + r)$$

and

$$R^2 - d^2 = 2Rr.$$

If we change the equation in the second part of the theorem to

$$R(R - 2r) = d^2,$$

we see that

$$R > 2r$$

for $d \neq 0$. We know that $d = 0$ occurs only when $\triangle ABC$ is an equilateral triangle. Therefore, we can say that if $\triangle ABC$ is not an equilateral triangle, then the radius of the circumscribed circle is greater than the diameter of the inscribed circle.

Euler's theorem is valid even if we change a triangle to a tetragon as indicated in the next theorem (see Figure 1.3).

THEOREM. *Given a tetragon $ABCD$, suppose that there exist the circumscribed circle $\widetilde{C}$ and the inscribed circle $\widetilde{C}'$. Then for any point P on $\widetilde{C}$, there exists a tetragon with P as one of the vertices such that $\widetilde{C}$ is the circumscribed circle about this tetragon and $\widetilde{C}'$ is the inscribed circle in this tetragon.*

If R and r are the radii of the circumscribed circle $\widetilde{C}$ and inscribed circle $\widetilde{C}'$, respectively, and d is the distance between the centers of the circles $\widetilde{C}$ and $\widetilde{C}'$, then

$$\frac{1}{(R+d)^2} + \frac{1}{(R-d)^2} = \frac{1}{r^2}.$$

These theorems are neither intuitive nor easy to prove. Indeed, some of the most fundamental features of this phenomenon, such as, for example, the inequality $R > 2r$ for nonequilateral triangles, would never manifest themselves. The equation describing the relation between R, r, and d (which indicates the relative locations of the circumscribed circle and the inscribed circle) is another example of the complexity and nonintuitive nature of these theorems.

♣ In general, for a given tetragon, there is neither a circumscribed circle about it nor a circle inscribed in it. Hence,

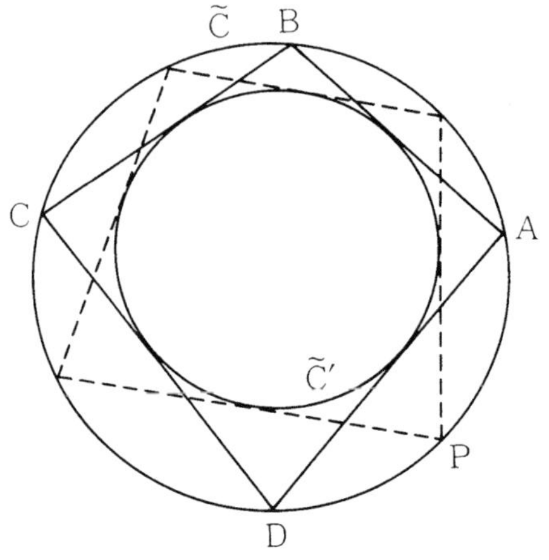

FIGURE 1.3

the tetragon which is mentioned in the above theorem is a very special one. However, once we have constructed such a tetragon, we can construct many more tetragons which have both circumscribed and inscribed circles.

1.2. The Poncelet Theorem

The Poncelet theorem is a generalization of the theorems discussed in the previous section. Poncelet extended the setting of the theorem from circles to ellipses, and from triangles and tetragons to general n-gons. By this "generalization", Poncelet elevated the theorem to a higher level.

THE PONCELET CLOSURE THEOREM I. *Suppose that it is possible to construct an n-gon inscribed in an ellipse C and circumscribed about an ellipse C'. Then, for any point P on C, there exists an n-gon with P as one of the vertices which is inscribed in C and circumscribed about C'.*

This theorem can be reformulated as follows.

THE PONCELET CLOSURE THEOREM II. *Given two ellipses C and C' where C' is located inside C, choose a point P on C. Starting at P, draw a line tangent to the inner ellipse C' and let P_1 be the point on C where this tangent line intersects the outer ellipse C. From P_1 draw another line tangent to the inner ellipse C' and let P_2 be the point on C where the second tangent line intersects the outer ellipse C. Repeat this process of drawing tangent lines to the inner ellipse C'. Let $P_{n-1}P_n$ be the segment of the tangent line to the inner ellipse C' after n applications of this process. If*

$$P_n = P \quad (n \geq 3),$$

then for any point P on the ellipse C, after n iterations of connecting points on C by lines tangent to C', the last intersection point will always be the same as the starting point (see Figure 1.4).

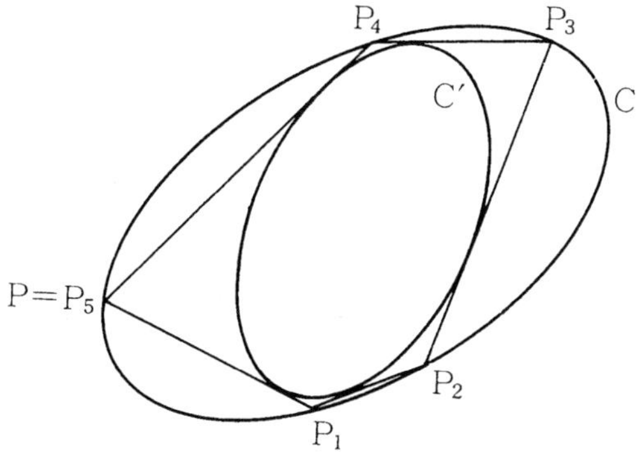

FIGURE 1.4

When C and C' are circles, several proofs are known. One such proof uses plane geometry, but it is rather difficult, so we will discuss it later in the chapter. However, if we replace the circles with ellipses, the existence of a proof using plane geometry becomes problematic. In fact, we need to use an algebraic method to prove the theorem. In working through this exercise, we will see how geometry and algebra interact.

To prove the above theorem we need to express the problem algebraically by using methods of analytic geometry and treating the relation between geometric figures as a relation between equations. This "translation" process consists of the following four steps.

(1) **Equations of the Ellipses C and C'.**

First we introduce the coordinate system by letting the origin be the center of the ellipse, the x-axis be the line along the major axis, and the y-axis be the line along the minor axis. Then we derive the following equation for the ellipse C:

$$\frac{x^2}{a^2} + \frac{y^2}{b^2} = 1 \quad (0 < b \le a). \tag{1}$$

In general, it is not possible to obtain a simple equation for the second ellipse C' using the same coordinate system. However, we can obtain the equation for C' from a standard equation of an ellipse $\widetilde{C}'$ centered at the origin by rotating the x-axis and y-axis and then translating the origin to a new position (β, γ). The equation of the ellipse centered at the origin is (see Figure 1.5):

$$\widetilde{C}' : \frac{x^2}{a'^2} + \frac{y^2}{b'^2} = 1. \tag{2}$$

If we rotate and translate $\widetilde{C}'$ we can bring $\widetilde{C}'$ to C'. This means we can obtain the equation for C' using the composition of a rotation and a translation of $\widetilde{C}'$.

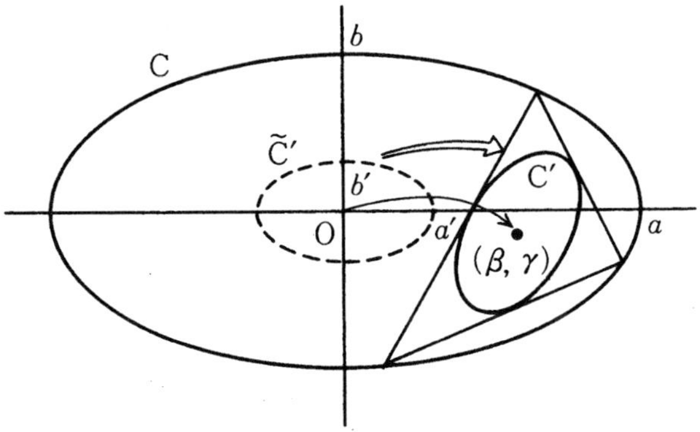

FIGURE 1.5

♣ If we substitute $x' = x\cos\theta + y\sin\theta$ and $y' = -x\sin\theta + y\cos\theta$ in the equation $\frac{x'^2}{a'^2} + \frac{y'^2}{b'^2} = 1$, we obtain the equation of the ellipse $\widetilde{C}'$ in the new coordinate system $x'y'$, where the axes are rotated about the origin by the angle θ. Furthermore, if we substitute $x - \beta$ for x and $y - \gamma$ for y, we obtain an equation of the ellipse with center translated to the point (β, γ).

Fortunately, to prove the Poncelet theorem, we do not need to have an exact equation for C'. It is sufficient to have any equation of the following form:

(3) $$\Phi(x, y) = Ax^2 + Bxy + Cy^2 + Dx + Ey + F = 0.$$

If $4AC - B^2 > 0$, then the above equation determines an ellipse. But the key point of the following proof is the fact that

From each point on C, we can draw two lines tangent to C'.

(2) **Parametric Representation of the Ellipse** C.

To discuss the proof in this section, we must first represent ellipses C and C' using parameters rather than the standard equations (1) and (3).

To obtain a parametric representation of the outer ellipse C, we first set

$$x = a\cos\theta \quad \text{and} \quad y = b\sin\theta \quad (0 \leq \theta < 2\pi)$$

(see Figure 1.6). Equations $x = a\cos\theta$, $y = b\sin\theta$ are called *parametric equations* of the ellipse C. The ellipse C is the set of points whose coordinates are given by the ordered pairs

(4) $$(a\cos\theta, b\sin\theta),$$

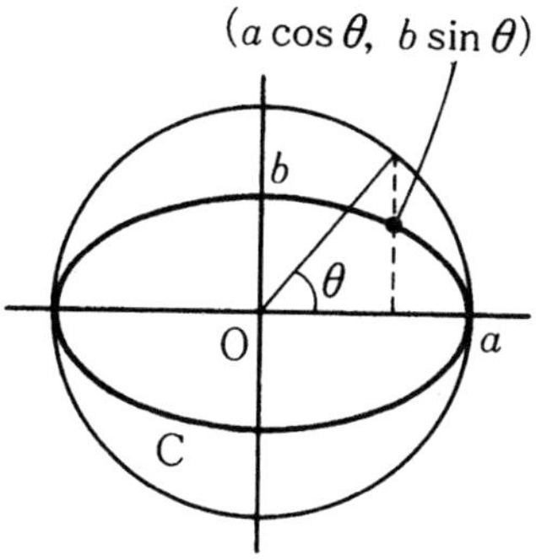

FIGURE 1.6

and an ordered pair (4) represents a point of the ellipse C with parameter θ. For our proof, we need to change the parameter θ to another parameter. We use the double angle formulas

$$\sin\theta = 2\sin\frac{\theta}{2}\cos\frac{\theta}{2},$$
$$\cos\theta = \cos^2\frac{\theta}{2} - \sin^2\frac{\theta}{2}$$

and rewrite the above equation as follows:

$$(5)\qquad \begin{aligned} \sin\theta &= \frac{2\sin\frac{\theta}{2}\cos\frac{\theta}{2}}{\sin^2\frac{\theta}{2}+\cos^2\frac{\theta}{2}} = \frac{2\tan\frac{\theta}{2}}{1+\tan^2\frac{\theta}{2}},\\ \cos\theta &= \frac{\cos^2\frac{\theta}{2}-\sin^2\frac{\theta}{2}}{\sin^2\frac{\theta}{2}+\cos^2\frac{\theta}{2}} = \frac{1-\tan^2\frac{\theta}{2}}{1+\tan^2\frac{\theta}{2}}. \end{aligned}$$

Let

$$(6)\qquad s = \tan\frac{\theta}{2}.$$

Since $0 \le \theta < 2\pi$, we have $0 \le \frac{\theta}{2} < \pi$. For $\theta \ne \pi$, the ranges in which the parameters θ and s vary are in one-to-one correspondence as indicated below (note that the case $\theta = \pi$ is an exception)

$$0 \le \theta < 2\pi(\theta \ne \pi) \Longleftrightarrow -\infty < s < \infty$$

(see Figure 1.7).

Substituting s for $\tan\frac{\theta}{2}$ in equation (5) and replacing $\cos\theta$ and $\sin\theta$ in ordered pairs (4) give us the following parametric representation for the outer ellipse C (except at the point $(-a, 0)$):

$$\left(\frac{a(1-s^2)}{1+s^2}, \frac{2bs}{1+s^2}\right) \qquad (-\infty < s < \infty).$$

We will regard the exceptional point $(-a, 0)$ as corresponding to the limit of the above parametric representation as s tends to $\pm\infty$.

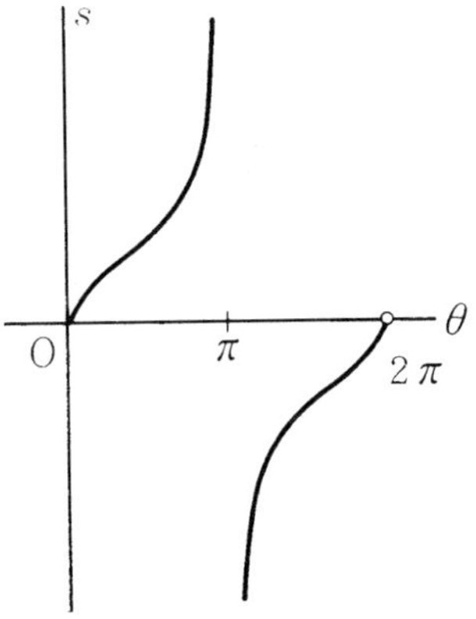

FIGURE 1.7

Finally by letting

$$x = f_1(s) = \frac{a(1-s^2)}{1+s^2}, \tag{7}$$
$$y = f_2(s) = \frac{2bs}{1+s^2}$$

we see that the points of the outer ellipse C can be represented as

$$(f_1(s), f_2(s)).$$

(3) **Parametric Representation of the Ellipse C'.**

We first obtain a parametric representation for the ellipse $\widetilde{C}'$ having the standard equation (2). By letting $t = \tan\frac{\theta}{2}$ (as in equation (6), but using t as our parameter), we have the following parametric representation for $\widetilde{C}'$ (except at the point $(-a', 0)$):

$$\left(\frac{a'(1-t^2)}{1+t^2}, \frac{2b't}{1+t^2}\right) \quad (-\infty < t < \infty).$$

We will regard the exceptional point $(-a', 0)$ as corresponding to the limit of the above parametric representation as s tends to $\pm\infty$.

Next we move $\widetilde{C}'$ to $\widetilde{C}$ by rotating and translating. Then the coordinates are transformed by a linear transformation and the addition of a constant. Therefore, we can say that any point on the ellipse C' can be represented by a parameter t as

$$\left(\frac{\widetilde{g}_1(t)}{1+t^2}, \frac{\widetilde{g}_2(t)}{1+t^2}\right) \quad (-\infty < t < \infty),$$

where

both $\widetilde{g}_1(t)$ and $\widetilde{g}_2(t)$ are quadratic polynomilas in t.

Now let

$$(8)\qquad \begin{aligned} x &= g_1(t) = \frac{\widetilde{g}_1(t)}{1+t^2}, \\ y &= g_2(t) = \frac{\widetilde{g}_2(t)}{1+t^2}. \end{aligned}$$

Taking into account equation (3) and the fact that (x, y) lies on C', we have

$$(9)\qquad \Phi(g_1(t), g_2(t)) = 0.$$

(4) **Equation for the Lines Tangent to C'.**

You may have previously studied the equation for a tangent line to an ellipse. If the ellipse is given by the standard equation

$$(10)\qquad F(x, y) = \frac{x^2}{a^2} + \frac{y^2}{b^2} - 1 = 0,$$

then the equation for the tangent line to this ellipse at a point (x_1, y_1) is given by

$$(11)\qquad \frac{xx_1}{A^2} + \frac{yy_1}{B^2} - 1 = 0.$$

♣ Equation (11) is obtained as follows. Consider first a line passing through the point (x_1, y_1) on the ellipse with slope m. Then, writing up the system of equations for this line and the ellipse, we obtain a quadratic equation for x. Next, from this quadratic equation we determine the value of m from the condition that this quadratic equation has x_1 as the double root. Equation (11) can also be obtained by differentiating equation (10) with respect to x, as follows:

$$\frac{2x}{A^2} + \frac{2y}{B^2}y' = 0.$$

Solving this equation for y', we obtain equation (11).

Notice that equation (11) gives us

$$\frac{x_1^2}{A^2} + \frac{y_1^2}{B^2} = 1.$$

Therefore, we can write the above equation as

$$\frac{2x_1}{A^2}(x - x_1) + \frac{2y_1}{B^2}(y - y_1) = 0.$$

The coefficient $\frac{2x_1}{A^2}$ of $(x - x_1)$ in the above equation is obtained by **differentiating the function $F(x, y)$ with respect to x, while regarding y as a constant**, and substituting $x = x_1$. The result is called the partial derivative of $F(x, y)$ with respect to x, and denoted by $F_x(x, y)$. By using this notation, we can write the coefficient $\frac{2x_1}{A^2}$ of $(x - x_1)$ as

$$F_x(x_1, y_1) = \frac{2x_1}{A^2}.$$

Similarly, differentiating $F(x, y)$ with respect to y while regarding x as a constant we obtain the partial derivative of $F(x, y)$ with respect to y. It is denoted by $F_y(x, y)$. Hence the coefficient $\frac{2y_1}{B^2}$ of $(y - y_1)$ can be written as

$$F_y(x_1, y_1) = \frac{2y_1}{B^2}.$$

Therefore, the equation for the tangent line to the ellipse at (x_1, y_1) is given by

$$F_x(x_1, y_1)(x - x_1) + F_y(x_1, y_1)(y - y_1) = 0.$$

If the ellipse C' is given by the equation (3),

$$\Phi(x, y) = Ax^2 + Bxy + Cy^2 + Dx + Ey + F = 0,$$

rather than by the standard equation, then the equation for the tangent line to ellipse C' at the point (x_1, y_1) on C' has the same form as above, i.e., the equation for the tangent tine is

$$\Phi_x(x_1, y_1)(x - x_1) + \Phi_y(x_1, y_1)(y - y_1) = 0. \tag{12}$$

We have now laid the groundwork for discussing the relation between the figures algebraically. Let points on the ellipses C and C' be expressed by parametric representations (7) and (8). For a point P on C with coordinates

$$P(f_1(s_0), f_2(s_0)),$$

let

$$Q_1(g_1(t_1), g_2(t_1))$$

be the point of intersection of C' and the line passing through the point P and tangent to C' (see Figure 1.8).

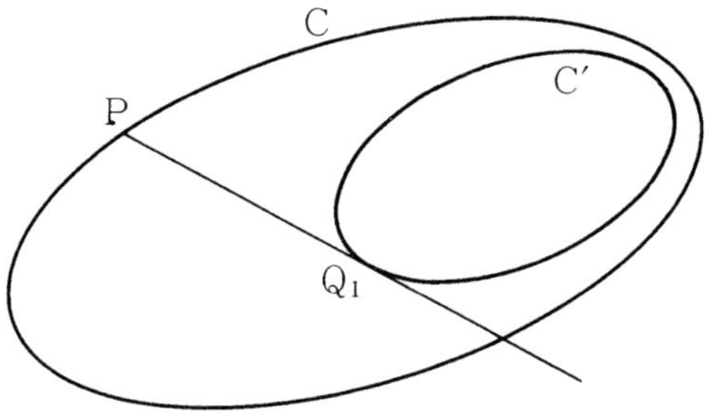

FIGURE 1.8

From equation (12), we know that the equation for the tangent line to C' at P is given by

$$\Phi_x(g_1(t_1), g_2(t_1))(x - g_1(t_1)) + \Phi_y(g_1(t_1), g_2(t_1))(y - g_2(t_1)) = 0.$$

Referring to equation (3) and actually calculating Φ_x and Φ_y, we see that the equation for the tangent line to P is given by

$$(2Ag_1(t_1)+Bg_2(t_1)+D)(x-g_1(t_1))+(Bg_1(t_1)+2Cg_2(t_1)+E)(y-g_2(t_1)) = 0.$$

Since the line represented by this equation must pass through the point $P(f_1(s_0), f_2(s_0))$, we have the equation

$$(13)\quad \begin{aligned} &(2Ag_1(t_1) + Bg_2t_1 + D)(f_1(s_0) - g_1(t_1)) \\ &\qquad + (Bg_1(t_1) + 2Cg_2(t_1) + E)(f_2(s_0) - g_2(t_1)) = 0. \end{aligned}$$

♣ As you can see, equation (13) is rather complicated, and you might not feel like generalizing it. However, if you do generalize this equation, you will encounter a phenomenon that will prove to be important to the following discussion. So let us now write out a generalization of the equation (13):

$$\begin{aligned} &(2Ag_1(t_1) + Bg_2(t_1) + D)f_1(s_0) \\ &\qquad + (Bg_1(t_1) + 2Cg_2(t_1) + E)f_2(s_0) \\ &\qquad - 2\Phi(g_1(t_1), g_2(t_1)) + Dg_1(t_1) + Eg_2(t_1) + 2F = 0. \end{aligned}$$

Taking into account equation (9), we have $\Phi(g_1(t_1), g_2(t_1)) = 0$. Hence, the term with this expression in the above equation disappears. This means that equation (13) does not contain terms that are quadratic with respect to g, such as $g_1(t_1)^2$, $g_2(t_1)^2$, or $g_1(t_1)g_2(t_1)$.

1.3. Introduction of the Algebraic Point of View

From now on, let $P(s_0)$ represent the point $P(f_1(s_0), f_2(s_0))$ on C and $Q(t_1)$ represent the point $Q(g_1(t_1), g_2(t_1))$. The fact that we can draw from $P(s_0)$ a tangent line to C' at $Q(t_1)$ led to a complicated relation. It seems hopeless at this point to try proving the Poncelet theorem using relation (13). However, the following fact about (13) can help us to prove the theorem. Referring to equations (7) and (8) we have

$$f_i(s) = \frac{\text{polynomial of second degree in } s}{1+s^2} \qquad (i = 1, 2),$$

$$g_i(t) = \frac{\text{polynomial of second degree in } t}{1+t^2} \qquad (i = 1, 2).$$

(Polynomials of second degree might degenerate into polynomials of first degree.) Hence, reducing (13) to common denominator, we obtain

$$\frac{Q(s_0, t_1)}{(1+s_0^2)(1+t_1^2)},$$

where $Q(s_0, t_1)$ is a polynomial in s_0 and t_1. Furthermore, we know that

$Q(s_0, t_1)$ is a polynomial of second degree in s_0 and in t_1.

To be a polynomial of second degree in s_0 means that if we view it as a polynomial in s_0 regarding t_1 as a constant, it has second degree (for example, (3) is a polynomial of second degree in x and in y).

Since $P(f_1(s_0), f_2(s_0))$ is an arbitrary point on C', we can replace s_0 by s and t_1 by t. Now let us look at the quadratic polynomial $Q(s,t)$ in s and t. If we let

$$Q(s,t) = 0,$$

we obtain an algebraic relation between s and t which expresses the existence of a tangent line to C' at t passing through s (see Figure 1.9).[1]

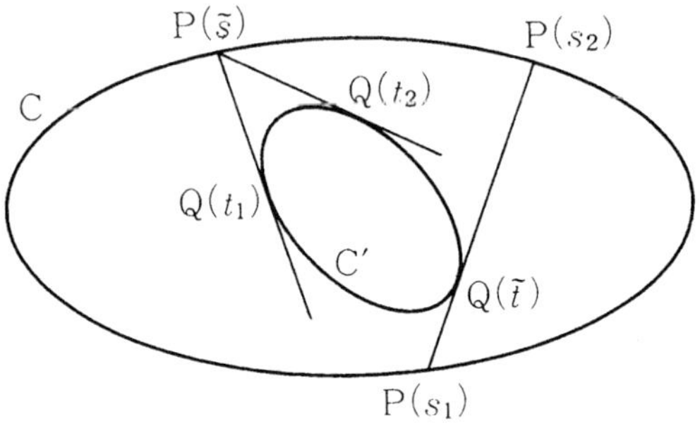

FIGURE 1.9

Therefore, if we view $Q(s,t) = 0$ as a quadratic equation in s, then one of the solutions of this equations is s_0. The other solution s_1 corresponds to the point of intersection $P(s_1)$ of C with the line tangent to C' at $Q(t_1)$ passing through $P(s_0)$ on C. That is, the relation between the two points $P(s_0)$ and $P(s_1)$ on C illustrated in Figure 1.10, is given by the *algebraic correspondence* via t_1 as follows:

$$Q(s_0, t_1) = 0 \quad \text{and} \quad Q(s_1, t_1) = 0.$$

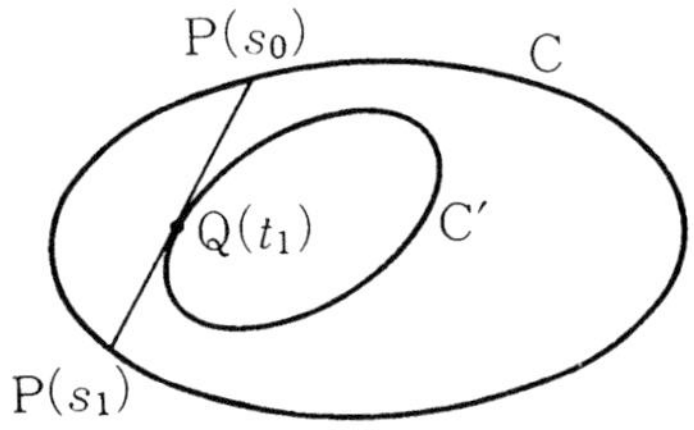

FIGURE 1.10

As is evident from Figure 1.10, once s_0 and s_1 are determined, t_1 should also be determined. We can generalize the above statement by saying that the relation between s_0 and s_1 is given by the equations

$$Q(s_0, t) = 0 \quad \text{and} \quad Q(s_1, t) = 0.$$

[1]This figure illustrates the relation between points on C and C'. For example, (1) If we choose $\tilde{s}$ corresponding to the point $P(\tilde{s})$ on C, then t_1 and t_2 corresponding to the points $Q(t_1)$ and $Q(t_2)$ on $\widetilde{C}$ are the solutions of the quadratic equation $Q(\tilde{s}, t) = 0$. (2) If we choose $\tilde{t}$ corresponding to the point $Q(\tilde{t})$ on $\widetilde{C}$, then s_1 and s_2 corresponding to the points $P(s_1)$ and $P(s_2)$ on C are the solution of the quadratic equation $Q(s, \tilde{t}) = 0$.

By eliminating t from these two equations, we obtain a relation

$$H(s_0, s_1) = 0$$

which gives the relation between s_0 and s_1.

$H(s_0, s_1)$ is an expression of second degree in s_0 and also in s_1.

We know that $H(s_0, s_1)$ is a quadratic polynomial in s_1 because the equation $H(s_0, s) = 0$ in s always has two solutions s_1 and s_1'. In this example, s_1' is the value of the parameter for the intersection point on C' with another tangent line to C' passing through $P(s_0)$ (see Figure 1.11). Similarly, $H(s_0, s_1)$ is an expression of degree 2 in s_0.

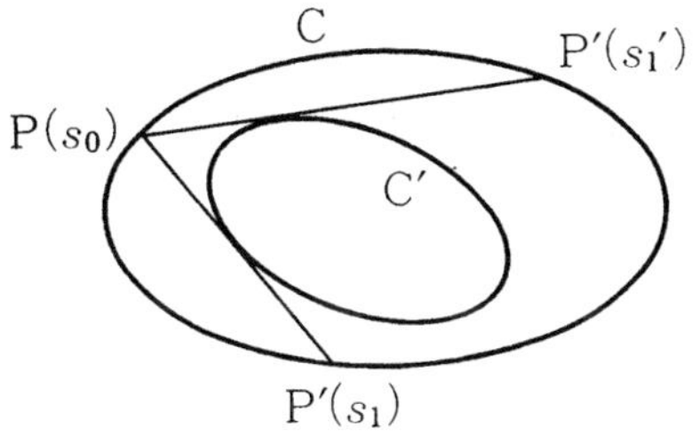

FIGURE 1.11

♣ Some of the readers might wonder if $H(s_0, s_1) = 0$ has complex solutions. Let us look at the case where s_0 is fixed. If the quadratic equation $H(s_0, s_1) = 0$ in s_1 had complex solutions, this would mean that from $P(s_0)$ we can draw a tangent line to C' in the "*complex* direction" (in the sense of *complex numbers*). As evidenced in Figure 1.11, we can draw two tangent lines to C' through $P(s_0)$ (the two tangent lines correspond to the two real solutions of $H(s_0, s_1) = 0$). If $H(s_0, s_1) = 0$ had imaginary solutions, it would be an equation of degree 3 or higher in s_1. Actually, when we have concluded before that $H(s_0, s_1)$ is an expression of degree 2 in s_1, we implicitly assumed that we cannot draw a tangent line in the "imaginary direction".

Some of you might feel puzzled that even though we are using real numbers, the world of complex numbers comes into play. Actually, the meaning of the Poncelet theorem becomes clearer if we consider it from the point of view of complex numbers. We will discuss this further in Lecture 3.

We previously stated that $H(s_0, s_1)$ is an expression of degree 2 in s_0 and also in s_1. This fact suggests an important point of view to look at the Poncelet theorem. Namely, starting from a point $P(s_0)$ on C and drawing a tangent line to C', we obtain another intersection point $P(s_1)$ on C. If we regard the relation

$$P(s_0) \longleftrightarrow P(s_1)$$

as a correspondence between two points on the ellipse C, we can say that this correspondence is induced by the inner ellipse C'. What we have shown is that this correspondence is expressed as the equation $H(s_0, s_1) = 0$, and it is an algebraic relation between two values of the parameter s.

Let us now revisit the Poncelet closure theorem II. In this theorem, an algebraic correspondence between two points on an ellipse is repeated several times, resulting in the sequence of points $P \to P_1 \to P_2 \to \cdots$. If we apply this process starting at the point P_1, and the last intersection point P_n coincides with the starting point after n applications, then the equation $P = P_n$ holds "identically" for each point P on C. The reason we use the word "identically" is because some of you might have already noticed that there is an algebraic aspect in the background of this theorem. In fact, the proof of the Poncelet theorem is accomplished with the help of the idea of algebraic correspondence.

♣ In 1878 Adolf Hurwitz (1859–1919) gave a proof using the notion of algebraic correspondence. We present his proof in the next section.

1.4. The Proof of the Poncelet Closure Theorem

Now, let us begin the actual proof! Take a point $P(s_0)$ of the ellipse C and draw the counterclockwise tangent line to the ellipse C' passing through $P(s_0)$. This tangent line intersects C again at the point $P(s_1)$ on C. Now, draw the tangent line to the ellipse C' passing through $P(s_1)$. This tangent line intersects C again at the point $P(s_2)$ on C. With this process, we obtain algebraic relations given by polynomials of degree 2 in s_0 and s_1, and s_1 and s_2,

$$H(s_0, s_1) = 0,$$
$$H(s_1, s_2) = 0.$$

If we eliminate s_1 from the above two relations, we obtain an algebraic relation between s_0 and s_2

$$H^{(2)}(s_0, s_2) = 0.$$

This relation is also described by a polynomial of degree 2 in s_0 and in s_2. This implies that the quadratic equation

$$H^{(2)}(s_0, s) = 0$$

in the variable s has only two solutions (see Figure 1.12).

Considering the correspondence between two points on C in the manner described in the Poncelet theorem, we obtain a sequence of correspondences

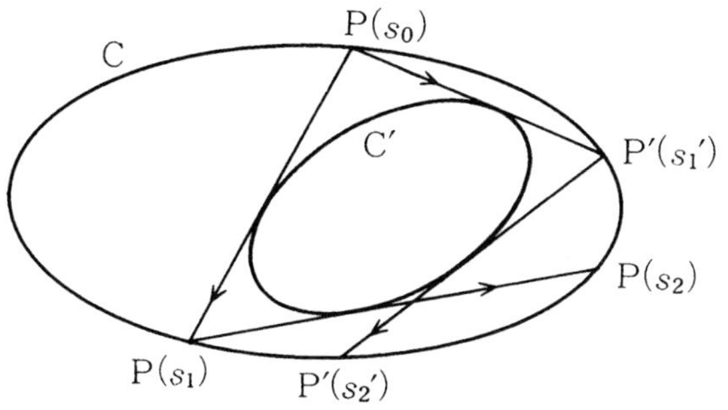

FIGURE 1.12

of points:

$$P(s_0) \to P(s_1) \to P(s_2) \to P(s_3) \to \cdots \to P(s_k) \to \cdots .$$

The algebraic relation between s_0 and s_3 is obtained from the relations described by the polynomials of degree two,

$$H^{(2)}(s_0, s_2) = 0,$$
$$H(s_2, s_3) = 0.$$

If we eliminate s_2 from the above two relations, we obtain an algebraic relation between s_0 and s_3,

$$H^{(3)}(s_0, s_3) = 0.$$

By continuing in this manner, we obtain an algebraic relation between s_0 and s_k as follows:

$$H^{(k)}(s_0, s_k) = 0.$$

Using the same arguments as before, we find that

$H^{(k)}(s_0, s_k) = 0$ is a relation of degree 2 in s_0 and in s_k.

Now, let us begin the proof of the theorem.

We will use the algebraic correspondence between points on the ellipse C that is induced by drawing a tangent line to the inner ellipse C'. Suppose we start at the point $P = P(s_0)$ and obtain the sequence of correspondences

$$P(s_0) \to P(s_1) \to P(s_2) \to \cdots \to P(s_{n-1}) \to P = P(s_0),$$

where the last point coincides with the starting point after n steps, $n \geq 3$. That is, the above algebraic correspondence is described by the relation

$$H^{(n)}(s_0, s_0) = 0.$$

Note that this is the algebraic formulation of the condition of the Poncelet closure theorem II. Now that you have a better feeling of what we are trying to prove, the mystery which so far seemed as impenetrable as the depth of the ocean, becomes somewhat clearer.

At this time we know that a hint to the proof is in the cyclical nature of the sequence of correspondences. For example, if we start at the point $P(s_1)$, we have

$$P(s_1) \to P(s_2) \to \cdots \to P(s_0)(= P) \to P(s_1),$$

where the last point corresponds to the starting point after n steps. Therefore, we have the relation

$$H^{(n)}(s_1, s_1) = 0.$$

Similarly, we have the relations

$$H^{(n)}(s_i, s_i) = 0 \quad (i = 0, 1, 2, \ldots, n-1).$$

It we consider the above relations from an algebraic point of view, they show that the quadratic equation

$$H^{(n)}(x, x) = 0$$

has n solutions s_0, s_1, s_2, ..., s_{n-1}. That is, the solutions of $H^{(n)}(s_j, x) = 0$ are the s_j obtained by the Poncelet sequence of correspondences

$$P(s_j) \to P(s_{j+1}) \to \cdots \to P(s_{n-1}) \to P(s_0) \to P(s_1) \to \cdots \to P(s_j),$$

and the counterclockwise construction of the sequence of correspondences

$$P(s_j) \to P(s_{j-1}) \to \cdots \to P(s_0) \to P(s_{n-1}) \to P(s_{n-2}) \to \cdots \to P(s_j).$$

Hence, the quadratic equation $H^{(n)}(x, x) = 0$ has $2n$ (≥ 6) solutions (counting with multiplicities). This can only happen when $H^{(n)}(x, x)$ is identically zero.

Therefore, for any point $\widetilde{P}(\tilde{s})$ on C we have

$$H^{(n)}(\tilde{s}, \tilde{s}) = 0.$$

The geometric interpretation of the above relation is that if we start at an arbitrary point $\widetilde{P}$ and apply the operation of drawing a tangent line to $\widetilde{C}$, then we will return to the starting point $\widetilde{P}$.

We have completed the proof of the Poncelet closure theorem.

♣ Jean Victor Poncelet (1788–1867) was born in northern France. He studied at the École Polytechnique in Paris. In early 1812, he joined the French army and served as a lieutenant of engineers in Napoleon's Russian campaign. In November of that year, Poncelet was captured by the Russian army and became a prisoner of war. He was placed in a prison camp in Saratov, on the Volga River, where he spent the next two years. During these two years of imprisonment, he studied projective properties of geometric figures without the use of any books. During this time, he established the fundamental theory of projective geometry. Poncelet's great work was published in Paris in 1822 as "Traité des propriétés projectives des figures". In sections 565–567 of this publication, you can find the theorem we have just discussed.

In his later years, Poncelet became the principal of the École Polytechnique and was involved in international exhibitions in London and Paris. He also devoted a great deal of time and effort to his administrative duties. In the last year of his life, he wrote an essay in which he said that he was unlucky because he could not devote his life to the study of geometry, which he considered to be his mission.

LECTURE 2

The Poncelet Theorem for Circles

2.1. The Case of Circles

In Lecture 1, we proved the Poncelet theorem in the general case of ellipses using an algebraic idea. It was necessary to modify our focusing on geometric figures by taking a "detour" through algebraic methods, which, while difficult, will make the rest of the journey easier! Hopefully, you felt some excitement as you followed the proof of the Poncelet theorem—perhaps one that you have not felt before in your study of mathematics. This is the thrill of solving enigmas of modern mathematics.

It may be of interest to note that the method we used to prove the Poncelet theorem has not been established until the late 19th century. However, Poncelet himself proved the theorem in the beginning of the 19th century using ideas from projective geometry. If the ellipses in the theorem happen to be circles, we can prove this theorem using different methods. A circle is geometrically simpler than an ellipse, and because of its simplicity, it also is a figure with many important properties. In fact, after Poncelet proved his theorem, Carl Gustav Jacob Jacobi (1804–1851) proved it using an analytic approach. Jacobi was a great mathematician of the early part of the nineteenth century. He showed that the Poncelet theorem for circles is somehow related to the elliptic integral mentioned in Chapter 3, Lecture 2. Jacobi proved the theorem using the idea of elliptic integrals. In this lecture we will discuss a simplified version of Jacobi's proof, given by Joseph Louis François Bertrand (1822–1900).

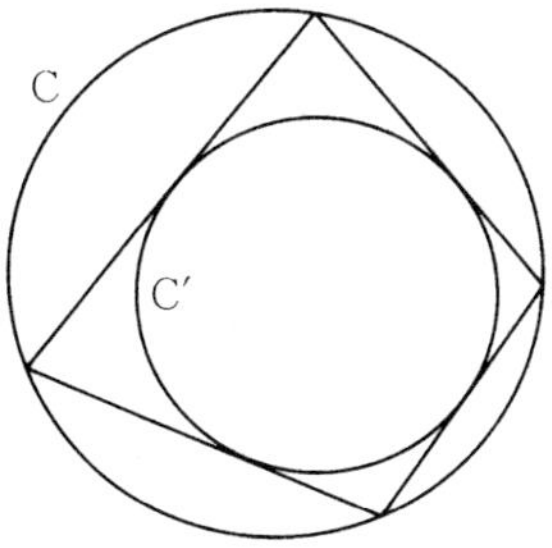

FIGURE 2.1

2.2. The Jacobian Invariant

Since the content of the Poncelet theorem is discussed extensively in Lecture 1, we shall not repeat it here. We consider two circles C and C' centered at O and O' with radii R and r, respectively. Let a (which will prove to be an important quantity) be the distance between the centers O and O' of these circles.

First, we explain Figure 2.2.

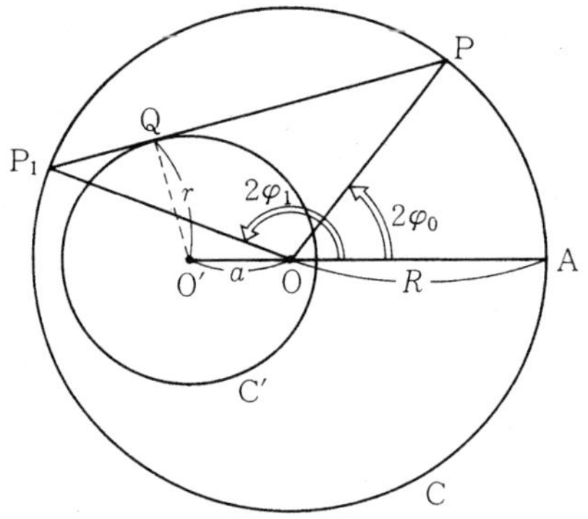

FIGURE 2.2

Choose a point P on C and draw a tangent line to C' passing through P, with Q being the point of tangency. This tangent line intersects C again at the point P_1 on C. We also set

$$\angle AOP = 2\varphi_0, \quad \angle AOP_1 = 2\varphi_1.$$

Jacobi discovered an astonishing fact about this figure: if $k > 0$ is determined by the formula

$$k^2 = \frac{4aR}{\sqrt{(R+a)^2 - r^2}},$$

then the value of the elliptic integral

$$\int_{\varphi_0}^{\varphi_1} \frac{d\varphi}{\sqrt{1 - k^2 \sin^2 \varphi}} \tag{1}$$

does not depend on the choice of P.

In other words, if we take another point $\widetilde{P}$ on C (Figure 2.3), and determine $2\tilde{\varphi}_0$ and $2\tilde{\varphi}_1$, then the value of the elliptic integral is the same as above, that is,

$$\int_{\varphi_0}^{\varphi_1} \frac{d\varphi}{\sqrt{1 - k^2 \sin^2 \varphi}} = \int_{\tilde{\varphi}_0}^{\tilde{\varphi}_1} \frac{d\varphi}{\sqrt{1 - k^2 \sin^2 \varphi}}. \tag{2}$$

We based the proof in Lecture 1 on an algebraic correspondence between P and P_1. We then expressed the relation between two points by the algebraic equation $H(s, s_0) = 0$. However, Jacobi looked at the relation

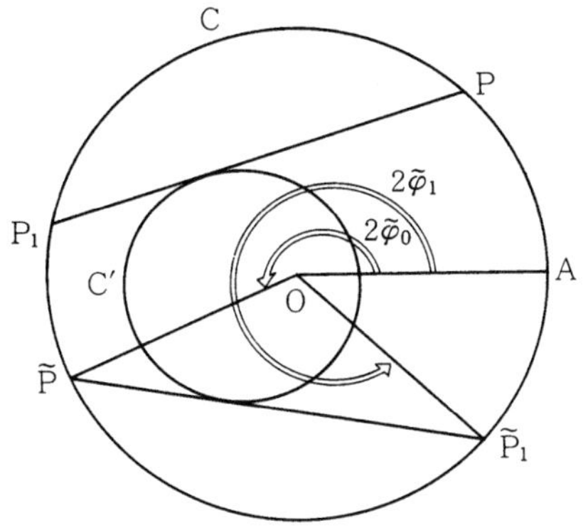

FIGURE 2.3

between two points from an analytic point of view, observing that if the angle 2φ is chosen as a parameter on C, then the relation between P and P_1 gives us the invariant quantity expressed as the integral (1). Now we are almost ready to prove equality (2). But first we prove the following result (see Figure 2.2):

$$QP = \sqrt{(R+a)^2 - r^2}\sqrt{1 - k^2 \sin^2 \varphi_0}, \tag{3}$$

$$QP_1 = \sqrt{(R+a)^2 - r^2}\sqrt{1 - k^2 \sin^2 \varphi_1}. \tag{4}$$

PROOF. Let $\triangle OO'P$ be the triangle with vertices O, O', and P as illustrated in Figure 2.4. By applying the second cosine formula to $\triangle OO'P$, we have

$$\begin{aligned}(O'P)^2 &= R^2 + a^2 + 2aR\cos 2\varphi_0 \\ &= (R^2 + a^2)(\cos^2\varphi_0 + \sin^2\varphi_0) + 2aR(\cos^2\varphi_0 - \sin^2\varphi_0) \\ &= (R+a)^2\cos^2\varphi_0 + (R-a)^2\sin^2\varphi_0 \\ &= (R+a)^2(1 - \sin^2\varphi_0) + (R-a)^2\sin^2\varphi_0.\end{aligned}$$

Therefore,

$$(O'P)^2 = (R+a)^2 - 4aR\sin^2\varphi_0.$$

Let $\triangle O'PQ$ be the right triangle shown in Figure 2.4. Applying the Pythagorean theorem to this triangle we have

$$\begin{aligned}QP &= \sqrt{(O'P)^2 - r^2} \\ &= \sqrt{(R+a)^2 - r^2 - 4aR\sin^2\varphi_0}.\end{aligned}$$

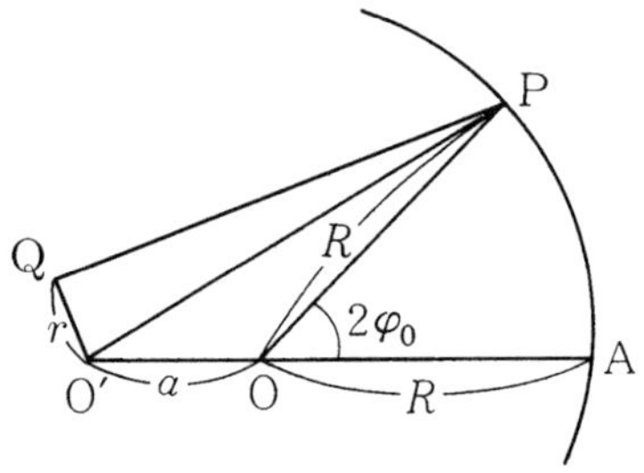

FIGURE 2.4

If we factor out $\sqrt{(R+a)^2 - r^2}$ from the right-hand side of the above equation and substitute

$$k^2 = \frac{4aR}{\sqrt{(R+a)^2 - r^2}}, \tag{5}$$

we obtain

$$QP = \sqrt{(R+a)^2 - r^2}\sqrt{1 - k^2 \sin^2 \varphi_0},$$

which is equation (3). We obtain equation (4) in a similar manner.

Thus, we proved equations (3) and (4). □

Once we choose a point P on C, we also determine P_1. Suppose a point on C is given by the angle measured from an the initial segment OA. Then we can say that once the angle $2\varphi_0$ is determined, the angle $2\varphi_1$ is also determined. In other words, φ_1 *is a function of* φ_0. The relation between φ_0 and φ_1 is expressed by a functional notation

$$\varphi_1 = \varphi_1(\varphi_0). \tag{6}$$

(The notation $\varphi_1 = f(\varphi_0)$ might be familiar to the reader. However, it is customary to use the more direct notation (6).) In Lecture 1, the point of view was that the correspondence between P and P_1 is an algebraic correspondence. In this lecture, we will change our point of view and regard this correspondence as a functional relation between φ_0 and φ_1. Then, we will use an analytic point of view. This is an important step allowing us to examine how an infinitesimal change of "the point" φ_0 results in the corresponding infinitesimal change of φ_1. That is, we consider the derivative of the function φ_1 in (6) with respect to φ_0 (see Figure 2.5).

Using the analytic point of view, we have the following result:

$$\frac{d\varphi_1}{d\varphi_0} = \frac{QP_1}{QP}. \tag{7}$$

PROOF. Let S be the point of intersection of the chords PP_1 and $P'P_1'$ of the circle C as shown in Figure 2.6.

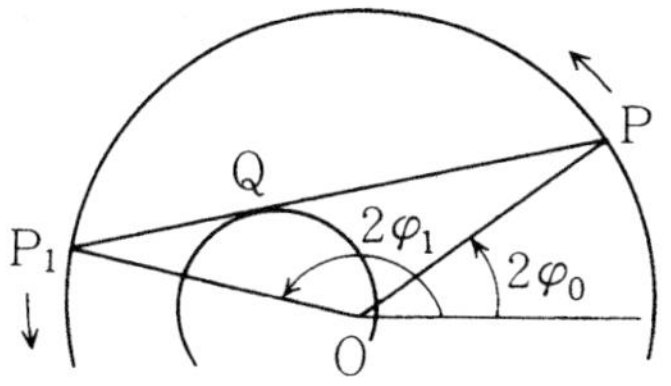

FIGURE 2.5. When P starts moving along the circle, the ratio of angular velocities of P and P_1, measured using φ_0 and φ_1, respectively, equals the ratio of the length of QP to the length of QP_1.

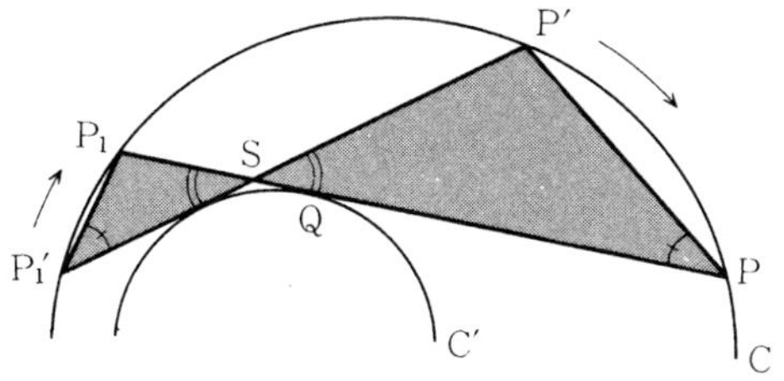

FIGURE 2.6

Note that

$$\triangle PP'S \propto \triangle P_1'P_1S \quad \text{(similar triangles)}.$$

We obtain this from the fact that $\angle P_1SP_1'$ and $\angle PSP'$ are congruent (since they are vertically opposite angles), as are the inscribed angles $\angle P$ and $\angle P_1'$, each based on the arc $P'P_1$.

Therefore,

$$\frac{P_1P_1'}{PP'} = \frac{SP_1}{SP'}.$$

As the point P' gets closer to P, the point P_1' gets closer to P_1—that is, $P_1' \to P$—and S gets closer to Q—that is, $S \to Q$—so that

$$\lim_{P' \to P} \frac{P_1P_1'}{PP'} = \frac{QP_1}{QP}.$$

On the other hand, if we choose P' sufficiently close to P and denote by $2\Delta\varphi_0$ the corresponding infinitesimal change of the angle, then for the arc length $\overset{\frown}{PP'}$ we have

$$\overset{\frown}{PP'} = R \cdot 2\Delta\varphi_0.$$

If $2\Delta\varphi_1$ is the infinitesimal change of the angle between P_1' and P_1, then

$$\overset{\frown}{P_1P_1'} = R \cdot 2\Delta\varphi_1.$$

Therefore,

$$\lim_{P' \to P} \frac{P_1 P_1'}{PP'} = \lim_{P' \to P} \frac{\widehat{P_1 P_1'}}{\widehat{PP'}} = \lim_{\Delta\varphi \to 0} \frac{\Delta\varphi_1}{\Delta\varphi_0} = \frac{d\varphi_1}{d\varphi_0},$$

which proves equation (7). □

By substituting (3) and (4) into equation (7), we obtain the following formula:

$$\frac{d\varphi_1}{d\varphi_0} = \frac{\sqrt{1 - k^2 \sin^2 \varphi_1}}{\sqrt{1 - k^2 \sin^2 \varphi_0}}. \tag{8}$$

Equation (8) gives us a clear picture of the correspondence between infinitesimal changes of φ_0 and φ_1. When the point P starts moving along the circle with angular velocity $\sqrt{1 - k^2 \sin^2 \varphi_0}$ (measured by φ_0), then the point P_1 starts moving along the circle with angular velocity $\sqrt{1 - k^2 \sin^2 \varphi_1}$ (measured by φ_1). What we have shown can be described as "analytically capturing the correspondence" between P and P_1.

Finally, we will introduce the **Jacobian Invariant**; that is, we are going to prove equation (2). To prove equation (2) we must show that the definite integral

$$\int_{\varphi_0}^{\varphi_1} \frac{d\varphi}{\sqrt{1 - k^2 \sin^2 \varphi}}$$

is a real number which does not depend on the choice of φ_0. Since φ_1 is determined by φ_0, the upper limit φ_1 in the integral can be written as $\varphi_1 = \varphi_1(\varphi_0)$. Hence, we denote

$$J(\varphi_0) = \int_{\varphi_0}^{\varphi_1(\varphi_0)} \frac{d\varphi}{\sqrt{1 - k^2 \sin^2 \varphi}}$$

and differentiate $J(\varphi_0)$ with respect to φ_0. After differentiation, we obtain

$$\frac{dJ}{d\varphi_0} = \frac{1}{\sqrt{1 - k^2 \sin^2 \varphi_1}} \frac{d\varphi_1}{d\varphi_0} - \frac{1}{\sqrt{1 - k^2 \sin^2 \varphi_0}}. \tag{9}$$

♣ For those of you who have already studied calculus: when we derive the fundamental theorem of calculus, we differentiate the following function:

$$F(x) = \int_a^x f(t)\, dt.$$

By differentiating both sides of this equation, we obtain

$$F'(x) = f(x).$$

Hence, if $x = x_1$, then $F'(x_1) = f(x_1)$. If, furthermore, x is a function of another variable s (that is, $x = x(s)$), then

$F(x) = F(x(s))$ (which is a composite function). This means that we can differentiate the function F with respect to s. By applying the chain rule for the differentiation of composite functions, we obtain

$$\frac{dF}{ds} = \frac{dF}{dx}\frac{dx}{ds} = f(x)\frac{dx}{ds}.$$

In the case of $x_1 = x_1(s_0)$, we have

$$F'(x_1(s_0)) = f(x_1)\frac{dx}{ds}(s_0).$$

Now, let

$$\widetilde{F}(x) = \int_s^{x(s)} f(t)\,dt.$$

To differentiate this function with respect to s, we choose a constant a and express the right-hand side as follows:

$$\widetilde{F}(x) = \int_a^{x(s)} f(t)\,dt - \int_a^s f(t)\,dt.$$

By differentiating the above function with respect to s, and evaluating it at $s = s_0$ and $x = x_1(s_0)$, we obtain

$$\frac{d\widetilde{F}}{ds}(s_0) = f(x_1)\frac{dx}{ds}(s_0) - f(s_0).$$

Substituting φ_1 for x_1 and φ_0 for s_0, we arrive at equation (9).

Substituting equation (8) into equation (9) we obtain

$$\frac{dJ}{d\varphi_0} = 0.$$

This result shows that the function $J(\varphi_0)$ is a constant, and so the value of the function does not depend on the choice of φ_0. Hence

$$J(\varphi_0) = J(\tilde{\varphi}_0)$$

for an arbitrary $\tilde{\varphi}_0$. Expressing the above equation in the integral form we have

$$\int_{\varphi_0}^{\varphi_1} \frac{d\varphi}{\sqrt{1 - k^2\sin^2\varphi}} = \int_{\tilde{\varphi}_0}^{\tilde{\varphi}_1} \frac{d\varphi}{\sqrt{1 - k^2\sin^2\varphi}}.$$

Thus, we have proved equation (2).

In the next lecture we will use the preceding material to prove the Poncelet theorem.

2.3. The Analytic Proof of the Poncelet Theorem

Let C and C' be circles of radii R and r, respectively, with C' lying inside C. Let a be the distance between the centers of the circles C and C'. Using equation (5), we can find the number k^2. Let us choose an arbitrary point P on C. Through that point, draw a tangent line to C'. This tangent line intersects C again at the point P_1. By drawing this tangent line we can determine the constant

$$\omega = \int_{\varphi_0}^{\varphi_1} \frac{d\varphi}{\sqrt{1-k^2\sin^2\varphi}}.$$

As we have just shown, ω does not depend on the choice of P.

Suppose we draw tangent lines to the inner circle C' described in the Poncelet theorem. If we repeat this process of drawing a tangent line k times, we will obtain the sequence of points

$$P \to P_1 \to P_2 \to \cdots \to P_k.$$

Let $2\varphi_0$ be the parameter corresponding to P, $2\varphi_1$ to P_1, and so on until we have $2\varphi_k$ corresponfing to P_k. Then

$$\begin{aligned}\int_{\varphi_0}^{\varphi_k} \frac{d\varphi}{\sqrt{1-k^2\sin^2\varphi}} &= \int_{\varphi_0}^{\varphi_1} \frac{d\varphi}{\sqrt{1-k^2\sin^2\varphi}} \\ &\quad + \int_{\varphi_1}^{\varphi_2} \frac{d\varphi}{\sqrt{1-k^2\sin^2\varphi}} + \cdots + \int_{\varphi_{k-1}}^{\varphi_k} \frac{d\varphi}{\sqrt{1-k^2\sin^2\varphi}} \\ &= \omega + \omega + \cdots + \omega \\ &= k\omega.\end{aligned}$$

If we repeat the process n times and the last point returns to the starting point so that $P_n = P$, we obtain

$$2\varphi_n = 2\varphi_0 + 2\pi.$$

Hence, we will have the following relation:

$$\int_{\varphi_0}^{\varphi_0+\pi} \frac{d\varphi}{\sqrt{1-k^2\sin^2\varphi}} = \int_{\varphi_0}^{\varphi_n} \frac{d\varphi}{\sqrt{1-k^2\sin^2\varphi}} = n\omega. \tag{10}$$

On the other hand, if the above relation holds, then we can say that $P = P_n$. That is, starting from P and drawing tangent lines to C n times, we go full circle and return to the starting point. Therefore, in order to prove the Poncelet theorem, we need to answer the following question.

> If equation (10) is valid for some φ_0, is it also valid for any other φ? That is, is the following statement true for any φ:
>
> $$\int_{\varphi}^{\varphi+\pi} \frac{d\varphi}{\sqrt{1-k^2\sin^2\varphi}} = n\omega?$$

The answer to this question is positive. This is because the integrand is a periodic function with period π. If a function is periodic, then the definite integral of that function over one period remains constant, no matter where the integration takes place.

♣ You can see that the integral of a periodic function over one period remains constant by studying Figure 2.7. The areas shown in Figure 2.7 are determined by the definite integrals (the area below the x axis is viewed as negative). The area over one period is the same regardless of the point we start.

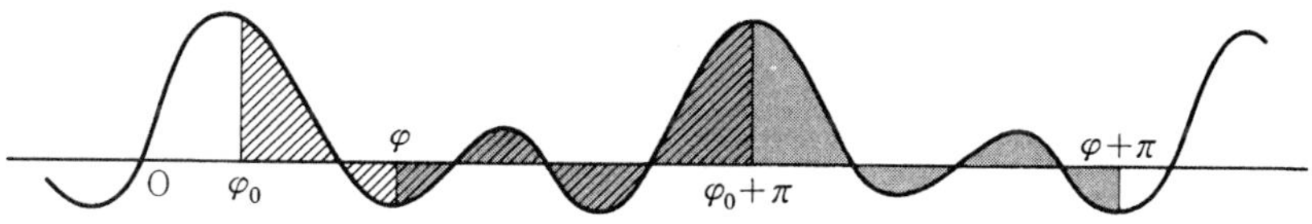

FIGURE 2.7

Therefore, for any φ we have

$$\int_{\varphi}^{\varphi+\pi} \frac{d\varphi}{\sqrt{1-k^2\sin^2\varphi}} = \int_{\varphi_0}^{\varphi_0+\pi} \frac{d\varphi}{\sqrt{1-k^2\sin^2\varphi}} = n\omega.$$

The above equation illustrates the case where the final point coincides with the starting point P. As a challenge we give the case where after m revolutions around the circle C the final point coincides with the starting point P. Thus, we have proved the Poncelet theorem.

CHALLENGE 2. If, starting from a point P on the circle C, we make a Poncelet-type construction with m revolutions so that the final point coincide with the starting point P, then starting from any other point on C, after m revolutions the final point will coincide with the starting point.

LECTURE 3

The Poncelet Theorem in the World of Complex Numbers

3.1. Introducing Complex Numbers

In Lectures 1 and 2, we proved the Poncelet closure theorem using only real numbers. For the limited purposes of these two lectures, we set aside the issue of complex numbers. In Lecture 2, where we were dealing exclusively with real analysis, the issue did not occur, while in Lecture 1, Section 1.4 we proved the Poncelet theorem algebraically under the assumption that there are no complex solutions to the given equations. This assumption was very important because implicit in it is the fact that in order to understand the Poncelet theorem more deeply, we must examine it in terms of complex numbers.

As you may recall, in the analytic proof of the Poncelet theorem for circles (based on the Jacobi's work), the notion of elliptic integrals was introduced. Gauss and Abel (we spoke about them briefly in Chapter 3), expanded the topic by considering the inverse functions of elliptic integrals in terms of complex numbers. Thus, even in the case of circles, complex numbers appear in the study of the Poncelet theorem.

In the context of plane geometry, a line and a circle intersect either at two points, or at a single point of tangency, or they do not intersect. As you can readily see, these various types of intersection give us the sense of variety and "flavor" of geometry. However, if we look at the intersection of a line and a circle from the point of view of algebraic equations, we find that they always intersect as long as we include complex numbers. The equation of a circle and of a line are

$$x^2 + y^2 = r^2,$$
$$y = mx + l,$$

respectively. If we eliminate y from this system, we obtain a quadratic equation for x,

$$x^2 + (mx + l)^2 = r^2.$$

This equation always has solutions as long as we allow complex numbers. In other words, if we expand the world of real numbers to the world of complex numbers, we can say that a circle and a line always intersect either at two points or at the point of tangency. Some of the points disappear from our physical view, but the solutions exist nonetheless. If the solutions are real numbers, we can physically see them as points of intersection.

By changing plane geometry problems into problems of algebraic equations using analytic geometry, mathematicians of the 19th century attempted to clarify the problems from a more general point of view. However, they realized that they needed to introduce complex numbers into their work. If we view the Poncelet theorem accordingly, what do we see? Let us give a simple example.

Until now, the ellipses or circles we discussed were such that one was located inside the other. In this lecture, however, we will consider the case of two intersecting circles C and C', where the center O' of C' is located on C.

Now we draw a tangent line to C' that passes through a point P on C, yet is outside of C' as shown in Figure 3.1. This tangent line intersects C again and let P_1 be this point of intersection. If we draw another tangent line to C' that passes through P_1, then this tangent line intersects C again and we denote by P_2 this point of intersection. From P_2, we will determine the point P_3 on C by a similar construction. This construction as illustrated in Figure 3.1 is slightly different from those we have seen previously. However, it is essentially Poncelet's construction with respect to two circles. If we apply the same construction used to determine P_3 to obtain P_4, we have $P_4 = P$.

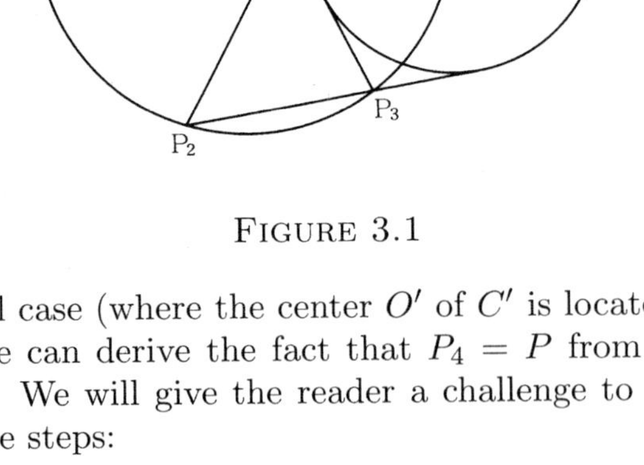

FIGURE 3.1

In the special case (where the center O' of C' is located on C as shown in Figure 3.1) we can derive the fact that $P_4 = P$ from the properties of inscribed angles. We will give the reader a challenge to prove this fact in the following nine steps:

CHALLENGE 3. By referring to Figure 3.2, verify the following nine facts.

(1) Show that the angles $\angle P_1Q'Q$, $\angle P_1QQ'$, $\angle P_1O'Q'$, and $\angle P_1O'Q$ are congruent (we call this angle α).

(2) If $\angle P = \beta$ in the right triangle $\triangle PQO'$, show that $2\alpha + \beta = 90°$.

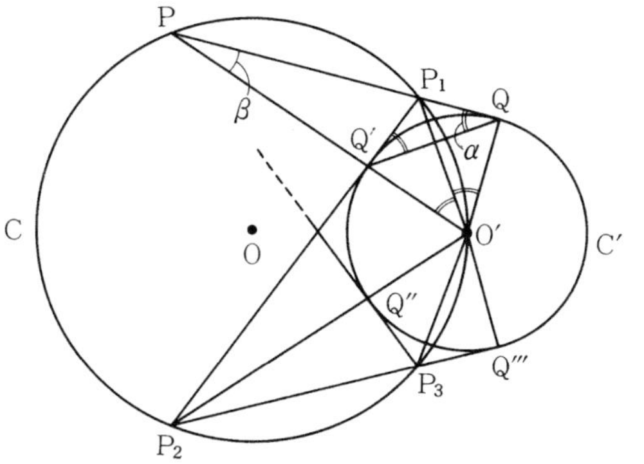

FIGURE 3.2

(3) Show that the intersection of the segment PO' and the circle C' is the point of tangency Q' of the tangent line drawn from P_1 to C'.

(4) Show that the intersection of the segment P_2O' and the circle C' is the point of tangency Q'' of the tangent line drawn from P_3 to C'.

(5) Show that $\angle Q'P_1O' = 90° - \alpha$.

(6) Show that $\angle P_2P_3O' = 90° + \alpha$ by using the relationship of the inscribed angles on the chord P_2O'.

(7) Show that $\angle P_3O'Q'' = \angle P_3O'P_2 = \alpha$.

(8) Show that the inscribed angle on the chord PO' of C which is on the side of O is equal to $90° - \alpha$.

(9) Show that P is on the line P_3Q''. (This implies that $P_4 = P$.)

Thus, in Challenge 3 we have proved that the Poncelet theorem is valid for two intersecting circles in the above special case. In this special case, we showed this by constructing four tangent lines with the final point coinciding with the starting point. This leads us to the question: "Is the Poncelet theorem valid for any two intersecting circles, or more generally, for any two intersecting ellipses?"

To answer this question, we need to explore this problem in more depth and, in particular, to extend the world of numbers, which is the background of this theorem, from real numbers to complex numbers.

Recall that in Lectures 1 and 2 we showed that it is possible to draw a tangent line to the inner circle O' from any point of the outer circle O. Now imagine that the inner circle O' was moved, so that now it intersects the outer circle O. You will see that it is impossible to draw a tangent line to the formerly inner circle O' from any point on the arc of O which lies inside O'. Therefore, somewhere in between the cases where one circle is entirely inside the other and only partially inside (or intersecting) the other, there is a gap, because in the first case we can always draw tangent lines and in the second case there are points from which we cannot draw tangent lines. However, if we treat these two cases as connected by a gradual continuous process, that is, if we continuously move O' from the initial position inside O to a position where it intersects O, then the perception of a gap disappears.

Perhaps the best way to visualize this concept of "gap" is to consider the quadratic equation

$$x^2 + ax + b = 0$$

and its graph.

The solutions of this equation are given as the points at which the graph of the function $y = x^2 + ax + b$ intersects the x-axis. Suppose the graph intersects the x-axis, but then, if b gets larger, the graph gradually moves upward and eventually will no longer intersect the x-axis (see Figure 3.3).

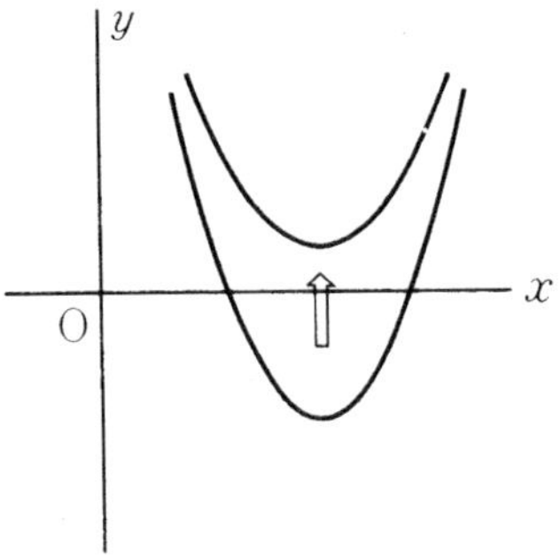

FIGURE 3.3

When the graph does not intersect the x-axis, there are no solutions of the equation as long as we confine ourselves to the world of real numbers. Even though the movement of the graph is continuous, we encounter a "gap" from having solutions to having no solutions. To get rid of this "gap" we need to expand the world of numbers from real to complex numbers. If we work within the complex number system, a quadratic equation always has solutions. If the coefficients of the equation vary continuously, the solutions change continuously as well. We may safely say that if we observe the graph of a function of degree 2 in the world of real numbers, the continuous correspondence between varying coefficients and solutions only partially comes into view on a single cross-section. As a result of limiting the point of view to real numbers, it appears that if there is no intersection, there are no solutions.

With the concept of a "gap" in mind we reexamine tangent lines to a circle once again. Let C be a circle of radius r,

$$C\colon x^2 + y^2 = r^2,$$

and let L be a line,

$$L\colon y = mx + l.$$

Let us then examine the points at which they intersect. We consider the case where the line L does not pass through the origin, that is, where $l \neq 0$.

(If $l = 0$, there are some special circumstances that will be discussed later in this section.)

The x-coordinates of the intersection points of C and L are obtained by eliminating y from the above system of equations, yielding the following result:

$$x^2 + (mx + l)^2 = r^2$$

or equivalently,

$$(1 + m^2)x^2 + 2lmx + l^2 - r^2 = 0. \tag{1}$$

By solving equation (1) for x we find the x-coordinates of the points at which C and L intersect. Now let D be the discriminant of this quadratic equation. Then

$$\begin{aligned}\frac{D}{4} &= l^2m^2 - (1 + m^2)(l^2 - r^2)\\ &= r^2(l + m^2) - l^2.\end{aligned}$$

If the discriminant D is positive, then the line L and the circle C intersect at two points. If $D = 0$, then the line L is tangent to the circle C. If D is negative, then the line L and the circle C do not intersect (see Figure 3.4). Since we are considering the solutions to the above equation only in the world of real numbers, here again, we encounter a "gap."

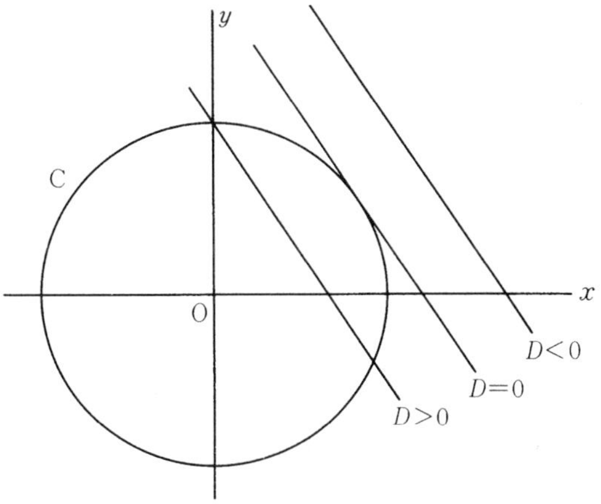

FIGURE 3.4

As you might already know, this "gap" happens because we have a graph which we can view physically, and attempt to examine the points of intersection in the world of real numbers. However, if we consider the solutions in the world of complex numbers, quadratic equation (1) always has solutions. They are given by the formula

$$x = \frac{-lm \pm \sqrt{r^2(1 + m^2) - l^2}}{1 + m^2}.$$

Hence, in the xy-coordinate plane where x and y are complex numbers, the points at which C and L intersect have the coordinates given by

$$\left(\frac{-lm \pm \sqrt{r^2(1+m^2)-l^2}}{1+m^2}, \frac{-lm^2 \pm m\sqrt{r^2(1+m^2)-l^2}}{1+m^2} + l \right).$$

Of course, if the discriminant D of quadratic equation (1) is negative, then the radical is expressed as

$$\sqrt{r^2(1+m^2)-l^2} = \sqrt{l^2 - r^2(1+m^2)}\, i,$$

which is a complex number.

Now let us once again briefly confine ourselves to the world of real numbers. If we consider the case where

$$0 < l < r,$$

then the line L cannot be a tangent line to C. This is because the intersection of the line $y = mx + l$ and the y-axis lies inside C since l is positive and less than r. This means that no matter what the slope m is, we cannot draw a tangent line to C through the point on the y-axis with the coordinate l (see Figure 3.5).

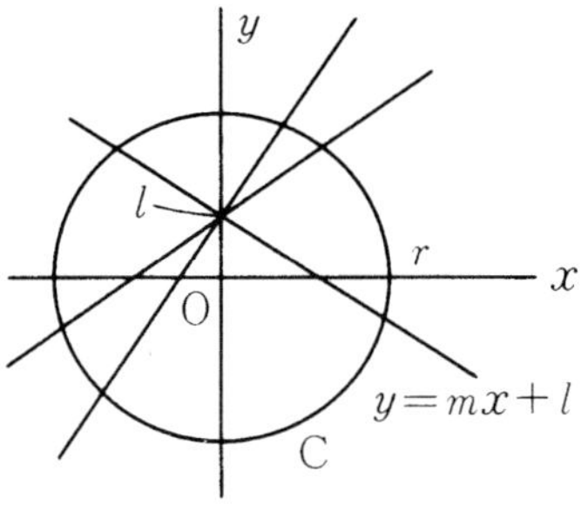

FIGURE 3.5

However, if we accept a broader point of view and work within the world of complex numbers, then we can have tangent lines even in the case $0 < l < r$. The condition for equation (1) to have multiple solutions is given by

$$\frac{D}{4} = r^2(1+m^2) - l^2 = 0.$$

We can rewrite this equation to obtain

$$\frac{l^2}{r^2} = 1 + m^2.$$

Since $0 < \frac{l}{r} < 1$, we have

$$-\left(1 - \frac{l^2}{r^2}\right) = m^2.$$

In other words, we have

$$m = \pm i\,\frac{\sqrt{r^2 - l^2}}{r}.$$

Hence, even lines that intersect the y-axis at the point l $(0 < l < r)$ can be viewed as tangents to C as long as they are drawn in the "complex" direction. These tangent lines are given by

$$y = \pm i\,\frac{\sqrt{r^2 - l^2}}{r}x + l. \tag{2}$$

In correspondence with the complex solutions of the quadratic equation in m, we have the idea of "complex tangent lines" to the circle C.

By introducing the idea of "complex" tangent lines, we can say that it is possible to draw two tangent lines to C' from any point on C, even if C and C' intersect. In Lecture 1, we proved the Poncelet theorem by using the correspondence between two points s_0 and s_1 on C obtained as the intersection points of the circle C and the tangent line to C'. Keeping this in mind, the idea of "complex" tangent lines to the circle C is therefore equivalent to considering the quadratic equation

$$H(s_0, s_1) = 0$$

not only in the world of real numbers, but also in the world of complex numbers. As you no doubt remember, the proof of the Poncelet theorem in Lecture 1 was algebraic, thus it would seem logical to apply a similar method of proof in the case of two intersecting circles. And, indeed, in this case the Poncelet theorem is valid.

If we look at the Poncelet Theorem from this "more advanced" point of view, which includes both real and complex numbers, then we can see the true nature of this theorem. Even so, a problem remains. Remember, we said previously that when we examine how the circle C and the line L intersect, there is a special case $l = 0$. This problem manifests itself when we examine the lines $y = mx$, which pass through the origin and intersect the circle C. If we apply equation (2) formally, then the lines passing through the origin and tangent to C are given by the equations

$$y = \pm ix.$$

The slopes m are $\pm i$, but if we substitute $y = \pm ix$ for y in the equation for the unit circle, we obtain

$$x^2 + y^2 = x^2 + (\pm ix)^2 = x^2 - x^2 = 0.$$

Hence, the lines $y = \pm ix$ do not have a common point with the circle $x^2 + y^2 = r^2$. So we have tangent lines to the circle that have no common point with the circle, which is curious indeed!

When $m = \pm i$, the coefficient $1 + m^2$ of x^2 in equation (1) equals 0. Such a strange phenomenon occurs because we can no longer use quadratic equations, which we have been using until now.

If we view the situation in which we can always draw two tangent lines to a given circle C from a point not on C as universal, then we can foresee that the setup must be expanded beyond the world of complex numbers. This expansion is provided by projective geometry. With projective geometry, we can see that the Poncelet theorem reaches a deeper aspect. We will discuss projective geometry in the next section. The work of Poncelet and Jacob Steiner (1796–1863) brought a significant expansion of geometry into different directions, which had a lasting impact on the 19th century mathematics. The other great direction of mathematics was established by Gauss, Riemann, and other 19th century mathematicians when they developed the elliptic function theory and algebraic function theory. Geometry had a significant influence upon all of the theories developed by these great mathematicians. In the 20th century, those theories, especially the elliptic function theory and the algebraic function theory, became the foundation of what we know today as algebraic geometry.

3.2. A Point at Infinity

From now on, the numbers we consider will be complex numbers. Therefore, the coefficients of an equation should not be restricted to real numbers only. The formula for the solutions of a quadratic equation in complex numbers has the same form as in the case of real numbers. This is because the formulas for factorization which are used to modify an equation are the same for complex numbers as for real numbers. Now consider a circle (as discussed in Section 3.1)

$$x^2 + y^2 = r^2.$$

(For convenience, let $r > 0$. Therefore, this is the standard equation for a circle of radius r.)

The equations for tangent lines to C passing through the origin, at least formally, are given by

$$y = \pm ix.$$

The points of tangency of those tangent lines to circle C do not exist in the xy-coordinate plane even if we consider x and y to be complex numbers. To find these points of tangency, we need to prepare a stage larger than the xy-coordinate plane.

> ♣ "Enlarging the stage," so to speak, is nothing new to mathematicians. That is precisely what they did when they extended the world of numbers in which we can solve given problems. For example, when people were only aware of natural numbers, they could not express the difference of 5 and 8 $(5 - 8)$. By adding zero and using the negative of natural numbers, they obtained integers. Using integers, the expression $5 - 8$ can be understood as -3. Similarly, people could not express the quotient of 3 divided by 5 by using integers. However, by extending the stage to include rational numbers,

we can express the quotient of 3 divided by 5 as $\frac{3}{5}$. Thus, we try to discover the missing intersection points of tangent lines $y = \pm ix$ and the circle C introducing "new" geometry by "enlarging the stage."

To enlarge the stage, we use the following idea: in the world of complex numbers, we introduce the coordinate transformation from xy-coordinates to uv-coordinates as follows:

$$\begin{cases} u = ix, \\ v = y. \end{cases}$$

Using this coordinate transformation the equations of a circle and tangent lines passing through the origin given by

$$\begin{cases} x^2 + y^2 = r^2, \\ y = \pm ix \end{cases} \tag{3}$$

will be changed accordingly:

$$\begin{cases} v^2 - u^2 = r^2, \\ v = \pm u. \end{cases}$$

If we restrict u and v to real numbers, these two equations represent a right-angle hyperbola and its asymptote, respectively, in the real uv-plane.

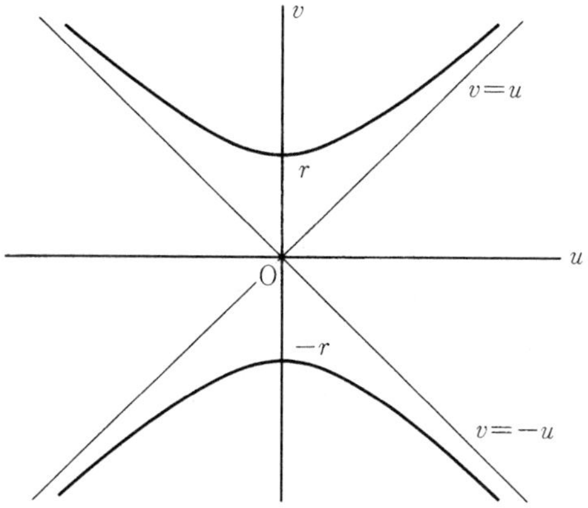

FIGURE 3.6

Strangely, most people find nothing wrong in the idea that the above hyperbola and each of its asymptotes intersect at the corresponding point at infinity. Even though we might not understand exactly where they intersect, the idea itself sounds reasonable. Hence, if we make another coordinate transformation and return the equations back to their original coordinate, it is natural to think that each intersection point of the circle and the tangent lines given by the system (3) is a point at infinity.

However, some of you might still feel a bit puzzled. A circle is a closed figure drawn in the finite range, so you might feel it cannot be stretched to infinity. Yet, tangent lines $y = \pm ix$ to the curves have points of tangency each at a point at infinity. But the circle given by

$$x^2 + y^2 = 1$$

appears as a finite figure only in the world of real numbers. If we extend the world of real numbers to that of complex numbers, then, for example, the point with complex coordinates

$$(10000, \sqrt{99999999}\, i)$$

satisfies the above equation of the unit circle. In other words, there exist points on the circle whose x-coordinate can be made as large as we want, and the points can approach infinity. In the world of complex numbers, the equation of a circle represents an open figure, just as a hyperbola does.

♣ Each point of tangency of the tangent lines $y = \pm ix$ to the circle $x^2 + y^2 = r^2$ is a point at infinity. This statement is based on the following facts. If we consider a line intersecting the y axis at the point l $(0 < l < r)$ and tangent to the circle $x^2 + y^2 = r^2$, then the x-coordinates of the points of tangency are given by

$$x = \mp ir\sqrt{\left(\frac{r}{l}\right)^2 - 1}.$$

We obtain these x-coordinates of the points of tangency using both the equations for the slope m and the solution to equation (1), which we discussed in the previous section. As l approaches 0, the absolute values of the above complex numbers increase. This shows that the closer the tangent line to the origin, the further away the points of tangency to the circle.

Even if we think that the discussed "points at infinity" are fictitious, the fact is that we have already started thinking about these "points at infinity" by developing some sort of geometric image in our minds. Girard Desargues (1591–1661) attempted to add "points at infinity" to the concept of a plane and of the space, and, in doing so, created a new geometry. Later in the 19th century, Poncelet, Steiner, and Julius Plücker (1801–1868) developed an exciting system of geometry known as projective geometry. By the beginning of the 20th century, the system of axioms for projective geometry had already been constructed.

♣ Desargues was a French army engineer, and an architect. Even though he was a practical man, he had a passion for pure mathematics. He was particularly interested in using the method of central projection for the study of conic sections. His most promising student was Blaise Pascal (who later would become very famous). In addition to the study of mathematics, Pascal later studied theology.

Let us briefly explain an important aspect of projective geometry. Consider two planes X and Y, made from transparent film. Now imagine a source of light P illuminating the planes X and Y, where X is positioned above Y (as in Figure 3.7). Then the point S shown in the figure will be projected to the point T on the plane Y.

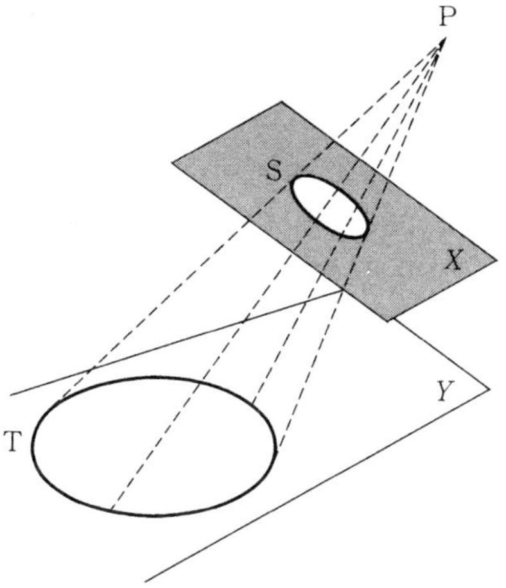

FIGURE 3.7. Projection of the plane X to the plane Y from the point P.

In this case we say that S is projected to T. In projective geometry, we identify properties that are common to both S and T and study them. For example, if the planes X and Y are parallel, then a triangle on X is projected onto Y as a similar but larger triangle.

However, if the planes X and Y are positioned as illustrated in Figure 3.8, then two parallel lines on the plane X are projected to two intersecting lines on the plane Y. In this case, a point at infinity, which can be reached by proceeding infinitely far along the parallel lines, is projected to the point on Y directly below the source of light P.

If an ellipse on the plane X is projected to the plane Y, the projected image changes from an ellipse to a parabola and then to a hyperbola as the ellipse moves along the plane X. If the ellipse on the plane X is placed entirely below (or above) the height of the point P, then the projected image on the plane Y is an ellipse. If we move the ellipse on the plane X upward until it reaches the height of the source of light P, we obtain a parabola on the plane Y. If we move the ellipse on the plane X further upward until the ellipse is positioned so that the top of the ellipse is above and the bottom of the ellipse is below the height of P, the image on the plane Y is a hyperbola (see Figure 3.9 and verify that we indeed obtain the required images).

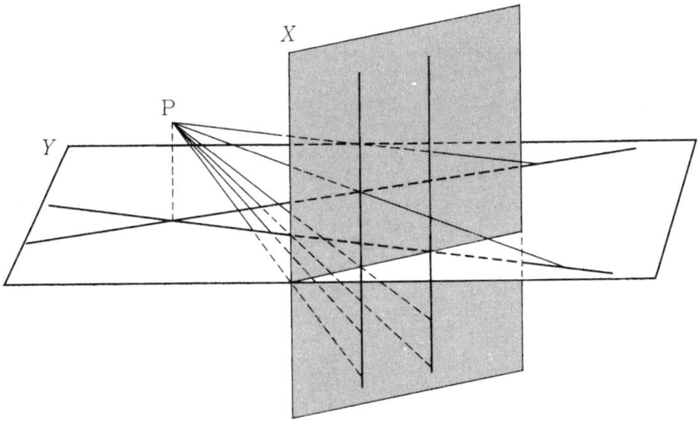

FIGURE 3.8. Two parallel lines on X are projected to two intersecting lines on Y.

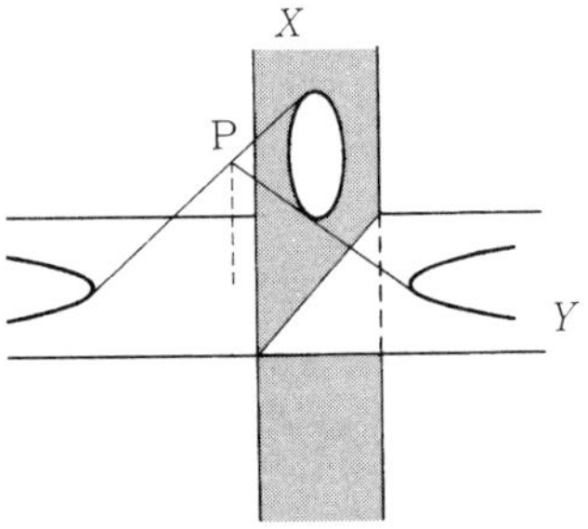

FIGURE 3.9. An ellipse on the plane X is projected to a hyperbola on the plane Y.

In projective geometry, the above phenomenon indicates that we can treat the three figures—ellipse, parabola, and hyperbola—as a curve which shares the same properties.

Without directly addressing projective geometry, we will now discuss a projective plane, with the above statement as a preliminary introduction. A projective plane in projective geometry has a role equivalent to that of the coordinate plane in Euclidean geometry. If we describe a projective plane in the world of real numbers, then the idea of projection we just discussed is intuitive. But we also discuss projective planes in the world of complex numbers. In doing so, we will develop a synthesis of the above discussions into a more meaningful system.

3.3. Projective Planes

We consider ordered triples of complex numbers (a_0, a_1, a_2) **except the triple** $a_0 = a_1 = a_2 = 0$. In other words, we consider triples where at least one component is nonzero. We say that two such triples (a_0, a_1, a_2) and

(b_0, b_1, b_2) have the same ratio, or are proportional to each other, if there exists $\alpha \neq 0$ such that

$$a_0 = \alpha b_0, \quad a_1 = \alpha b_1, \quad \text{and} \quad a_2 = \alpha b_2.$$

We denote this relation by

$$a_0 : a_1 : a_2 = b_0 : b_1 : b_2.$$

CHALLENGE 4. (i) If a_0, a_1, and a_2 are all nonzero, show that

$$a_0 : a_1 : a_2 = 1 : \frac{a_1}{a_0} : \frac{a_2}{a_0} = \frac{a_0}{a_1} : 1 : \frac{a_2}{a_1} = \frac{a_0}{a_2} : \frac{a_1}{a_2} : 1$$

(ii) Show that the condition $a_0 \neq 0$ is equivalent to the statement

$$a_0 : a_1 : a_2 = 1 : b_1 : b_2$$

for some b_1 and b_2.

Show also that in this case b_1 and b_2 are uniquely determined by a_0, a_1 and a_2.

Now we consider the ratios $x_0 : x_1 : x_2$ of all ordered triples of complex numbers $(x_0, x_1, x_2) \neq (0, 0, 0)$. The set of such ratios is called the **complex projective plane** and is denoted by $\mathbb{P}^2(\mathbb{C})$. We can regard these ratios as coordinates of points in $\mathbb{P}^2(\mathbb{C})$. We say that a point P in the complex projective plane has coordinates $(x_0 : x_1 : x_2)$. The symbol $P(x_0, x_1, x_2)$ also will be used to represent the point P. Since the ratios of coordinates of two propotional triples coincide, we have

$$P(x_0 : x_1 : x_2) = P(\alpha x_0 : \alpha x_1 : \alpha x_2)$$

for any α ($\neq 0$). In the case where $x_0 \neq 0$, we can express P as

$$P(x_0 : x_1 : x_2) = P\left(1 : \frac{x_1}{x_0} : \frac{x_2}{x_0}\right). \tag{4}$$

Conversely, every pair of complex numbers x and y determines a point

$$P(1 : x : y)$$

in $\mathbb{P}^2(\mathbb{C})$ satisfying the condition $x_0 \neq 0$. (See Challenge 4, problem (ii)).

We pay special attention to points $P(1 : x : y)$ in $\mathbb{P}^2(\mathbb{C})$ and consider such a point to be the same as the point with coordinates (x, y) in the xy-plane. In other words, a point (x, y) in the plane $\mathbb{C}^2$ is considered to represent the point $(1 : x : y)$ in $\mathbb{P}^2(\mathbb{C})$, i.e., we associate with each point (x, y) in $\mathbb{C}^2$ the point $(1 : x : y)$:

$$(x, y) \in \mathbb{C}^2 \to (1 : x : y) \in \mathbb{P}^2(\mathbb{C}).$$

♣ The symbol $\mathbb{C}^2$ on the left-hand side of the arrow denotes the complex coordinate plane in which a point is represented by an ordered pair of complex numbers as coordinates. The symbol $\mathbb{P}^2(\mathbb{C})$ denotes the complex projective plane, with $\mathbb{P}$ indicating that it is a projective plane and $\mathbb{C}$ indicating that complex numbers are used. You might wonder why, even though we talk about ratios of three numbers, we use the superscript 2. The reason is that only two values are determined by the ratio of three numbers.

Using the above correspondence between the points of $\mathbb{C}^2$ and $\mathbb{P}^2(\mathbb{C})$, we can view the complex coordinate plane $\mathbb{C}^2$ as sitting inside $\mathbb{P}^2(\mathbb{C})$, that is,

$$\mathbb{C}^2 \subset \mathbb{P}^2(\mathbb{C}).$$

We can see that a familiar coordinate plane (in this case, it is a complex coordinate plane) is contained in a "larger plane" $\mathbb{P}^2(\mathbb{C})$. What we need to ask now is which points are missing from, or newly adjoined to, $\mathbb{C}^2$. The answer is that they are the points represented by the triples

$$(0 : x_1 : x_2).$$

We call a point represented by the above notation a **point at infinity** and the set of all such points the **line at infinity**. We denote the line at infinity by l_∞. With this new terminology, the complex coordinate plane $\mathbb{C}^2$ together with the line at infinity l_∞ is the complex projective plane $\mathbb{P}^2(\mathbb{C})$.

♣ In Chapter 2, Lecture 3, Section 3.2, we discussed points at infinity in our materials on general three-dimensional figures. We introduced the point at infinity while creating a sphere from a plane by adding a point to it. That point at infinity is different from the point at infinity discussed in this lecture. In this lecture, points at infinity lie on a line which is located at infinity. When we observe infinity from a finite distance, we tend to see it as something lying far away. We conjure up myriad images to help us to grasp the idea. Similarly, in mathematics, there are many ways to introduce the concept of a point at infinity, depending upon the object being considered.

Here we have used the expression "at infinity." So let us now explore the delicate meanings of this expression by using two parallel lines as an example. In order to develop a better understanding of parallel lines, consider first two parallel lines in the plane $\mathbb{C}^2$:

$$\begin{aligned} y &= mx + \alpha, \\ y &= mx + \beta \qquad (\alpha \neq \beta). \end{aligned}$$

Referring to expression (4), if we consider these two parallel lines in the complex projective plane, the equations for the two lines may be written as

$$\begin{aligned} \frac{x_2}{x_0} &= m\frac{x_1}{x_0} + \alpha, \\ \frac{x_2}{x_0} &= m\frac{x_1}{x_0} + \beta. \end{aligned}$$

Multiplying both sides by x_0 and simplifying the resulting equations we obtain

$$\alpha x_0 + m x_1 - x_2 = 0, \tag{5}$$

$$\beta x_0 + m x_1 - x_2 = 0. \tag{6}$$

These equations are regarded as representing two lines in $\mathbb{P}^2(\mathbb{C})$.

♣ We define a line in a projective plane as follows: if for some constants a, b, c with at least one of them being nonzero, three numbers x_0, x_1, and x_2 satisfy the relation

$$a x_0 + b x_1 + c x_2 = 0,$$

then we say that the point $(x_0 : x_1 : x_2)$ is on the line in $\mathbb{P}^2(\mathbb{C})$. Notice that the above equation is of first degree in x_0, x_1, and x_2. Also, if those three numbers satisfy the equation, then so do αx_0, αx_1, αx_2.

Now let us determine the point of intersection of the two parallel lines (5) and (6) by subtracting equation (6) from equation (5):

$$(\alpha - \beta) x_0 = 0.$$

Since $\alpha \neq \beta$, we get

$$x_0 = 0.$$

From equation (5), we have

$$x_2 = m x_1.$$

Hence, we can conclude that the point of intersection of the parallel lines (5) and (6) is

$$(0 : 1 : m).$$

The coordinates of this point of intersection indicate that it is on the line at infinity l_∞. We have shown that two parallel lines in $\mathbb{P}^2(\mathbb{C})$ intersect at the point at infinity $(0 : 1 : m)$.

We can state this differently: a line with the slope m,

$$y = mx + \alpha,$$

in $\mathbb{P}^2(\mathbb{C})$ always intersects the line at infinity at the point $(0 : 1 : m)$. On the other hand, the line

$$x = \alpha$$

can be written as $\alpha x_0 - x_1 = 0$. We know that this line intersects the line at infinity $x_0 = 0$ at $(0 : 0 : 1)$.

As you can see, each point $(0 : a : b)$ on the line at infinity corresponds to the slope $m = \frac{b}{a}$ of the line $y = mx + \alpha$. If a line passing through the origin is given by the equation

$$bx - ay = 0,$$

then it intersects the line at infinity at the point $(0 : a : b)$ (see Figure 3.11).

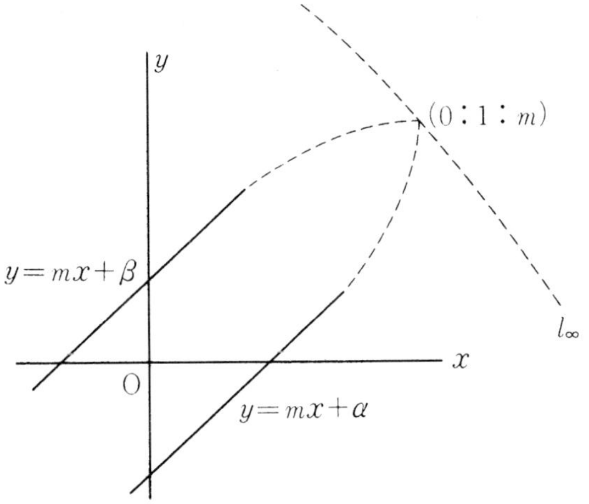

FIGURE 3.10

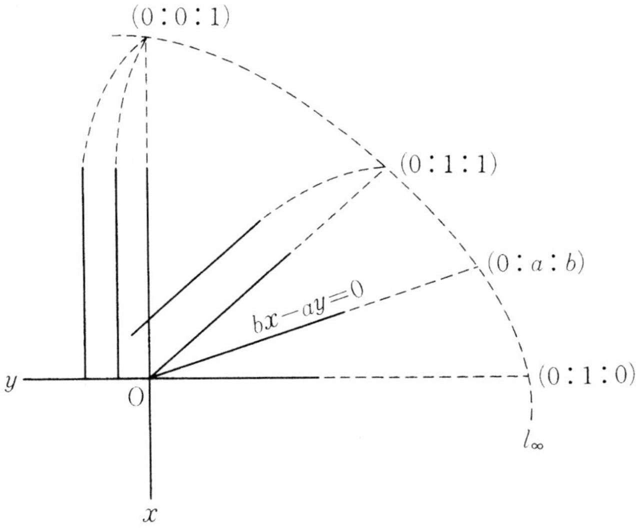

FIGURE 3.11

By this time, you hopefully have a stronger grasp on the notion of a point at infinity on a line and a line at infinity in a plane.

Next we consider the hyperbola

$$\frac{x^2}{a^2} - \frac{y^2}{b^2} = 1 \quad (0 < a < b) \tag{7}$$

in a complex projective plane. The asymptotes for hyperbola (7) are given by

$$y = \pm\frac{b}{a}x.$$

As we have just seen, these asymptotes pass through the points at infinity $(0 : a : \pm b)$.

The larger $|x|$ gets, the closer the hyperbola approaches the asymptotes. Therefore, some readers might have guessed that hyperbola (7) also passes through the points $(0 : a : \pm b)$. This is indeed the case. In fact, the hyperbola and its asymptotes intersect at points at infinity. Let us further examine this fact.

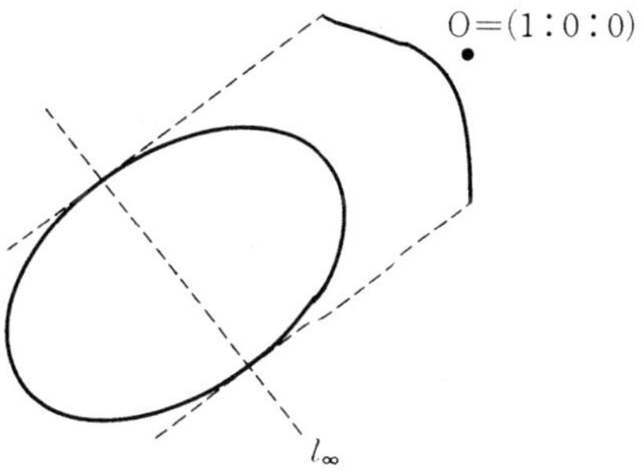

FIGURE 3.12. The hyperbola and its asymptotes viewed from the line at infinity.

In $\mathbb{P}^2(\mathbb{C})$, the hyperbola given in (7) is expressed as

$$\frac{x_1^2}{a_2} - \frac{x_2^2}{b^2} - x_0^2 = 0.$$

From this equation, we can readily see that the hyperbola passes through the points $(0 : a : \pm b)$. To find the points of intersection of the hyperbola and its asymptote $y = \frac{b}{a}x$, we must solve the system of equations

$$\begin{cases} x_2 - \dfrac{b}{a}x_1 = 0, \\ \dfrac{x_1^2}{a^2} - \dfrac{x_2^2}{b^2} - x_0^2 = 0. \end{cases}$$

Substituting $x_2 = \frac{b}{a}x_1$ for x_2 in the second equation in the system, we obtain

$$x_0^2 = 0.$$

Therefore, the point of intersection is $(0 : a : b)$, and it is a "double point" since $x_0^2 = 0$. This means that the asymptote is tangent to the hyperbola at the point of intersection.

Similarly, consider the parabola $y^2 = x$. In $\mathbb{P}^2(\mathbb{C})$, this parabola is given by the equation

$$x_0x_1 - x_2^2 = 0.$$

So we know that the parabola passes through the point at infinity $(0 : 1 : 0)$. Since the intersection of the parabola and the line at infinity satisfies the condition $x_2^2 = 0$, we also know that the line at infinity is tangent to the parabola at $(0 : 1 : 0)$.

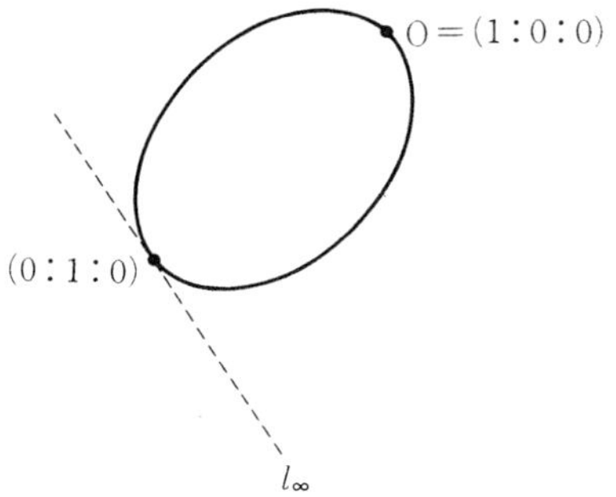

FIGURE 3.13. The parabola viewed from the line at infinity.

If we confine ourselves to the xy-plane $\mathbb{C}^2$, the points at infinity or the line at infinity lie beyond the horizon. However, if we broaden our point of view to $\mathbb{P}^2(\mathbb{C})$, the point at infinity and the line at infinity are not problematic locations anymore. They are ordinary points and an ordinary line in $\mathbb{P}^2(\mathbb{C})$. This is a rather important fact, so let us emphasize it as follows: suppose a line in $\mathbb{P}^2(\mathbb{C})$ is given by $ax_0 + bx_1 + cx_2 = 0$. If $a = 1$ and $b = c = 0$, then it represents the line at infinity $x_0 = 0$. If $a = c = 0$ and $b = 1$, then it represents the line $x_1 = 0$, which is the y-axis if it is observed in the plane $\mathbb{C}^2$. On the other hand, if $a = b = 0$ and $c = 1$, then the line becomes $x_2 = 0$, which is the x-axis if it is observed in the plane $\mathbb{C}^2$. In short, the x-axis, y-axis, and the line at infinity all have the same form when observed in $\mathbb{P}^2(\mathbb{C})$.

The circle $x^2 + y^2 = 1$ is expressed as the curve $x_0^2 - x_1^2 - x_2^2 = 0$ in $\mathbb{P}^2(\mathbb{C})$ and it intersects the line at infinity at $(0, 1, \pm i)$.

♣ As long as we confine ourselves to the world of real numbers, circles and ellipses are finite (closed). However, if we "move" these circles and ellipses so that each of them intersects the line at infinity at one point, they become parabolas. If we "move" them even further so that each of them will intersect the line at infinity at two distinct points, they become hyperbolas.

In $\mathbb{P}^2(\mathbb{C})$ we can no longer see, as we did in the case of real numbers, the difference between ellipses, parabolas, and hyperbolas. Now this difference manifests itself in the relative position of the curves and the line at infinity. So, we view these curves as mathematical equations and consider them to be equivalent.

This point of view is mathematically established as follows. Let

$$F(x_0, x_1, x_2)$$

be an irreducible (i.e., one that cannot be factorized) homogeneous polynomial of second degree. Then the set of all zero points $x_0 : x_1 : x_2$ of the homogeneous equation $F(x_0, x_1, x_2) = 0$ is called an irreducible plane curve of degree 2. The polynomial $F(x_0, x_1, x_2)$ is expressed as

$$F(x_0, x_1, x_2) = \sum_{\substack{i_0+i_1+i_2=2 \\ i_0, i_1, i_2 > 0}} x_0^{i_0} x_1^{i_1} x_2^{i_2}.$$

(This sum is over all indices i_0, i_1, i_2 such that $i_0+i_1+i_2 = 2$ and $i_0, i_1, i_2 \geq 0$.)

In $\mathbb{C}^2$, coordinate transformations $(x, y) \to (x', y')$ are represented by linear transformations which leave the length unchanged. In other words, lengths are invariant under those transformations. In $\mathbb{P}^2(\mathbb{C})$, coordinate transformations $(x_0 : x_1 : x_2) \to (x_0' : x_1' : x_2')$ are given by the following relations:

$$\rho x_0' = a_{00}x_0 + a_{01}x_1 + a_{02}x_2,$$

$$\rho x_1' = a_{10}x_0 + a_{11}x_1 + a_{12}x_2,$$

$$\rho x_2' = a_{20}x_0 + a_{21}x_1 + a_{22}x_2,$$

where ρ is a nonzero constant and the determinant satisfies the following condition:

$$\begin{vmatrix} a_{00} & a_{01} & a_{02} \\ a_{10} & a_{11} & a_{12} \\ a_{20} & a_{21} & a_{22} \end{vmatrix} \neq 0.$$

Such transformations are called **projective transformations**.

It is known that by a certain projective transformation we can always reduce the irreducible polynomial $F(x_0, x_1, x_2)$ to the polynomial

$$-x_0'^2 + x_1'^2 + x_2'^2 = 0.$$

Therefore, each irreducible curve of degree 2 can be identified with just one plane curve

$$x_0^2 - x_1^2 - x_2^2 = 0.$$

If we view this plane curve of degree 2 in the xy-plane, it is just the unit circle

$$x^2 + y^2 = 1.$$

3.4. Tangent Lines to Plane Curves of Degree 2

While using an algebraic method to prove the Poncelet closure theorem, the following question arose: "How many tangent lines to an ellipse (and, in general, to a plane curve of degree 2) can be drawn from a point P not on the ellipse (or the plane curve)?" The answer to this question is found in the projective plane $\mathbb{P}^2(\mathbb{C})$. To a given curve of degree 2, exactly two tangent lines can be drawn from any such point P. Once you understand this fact, which we will discuss further in this section, you will see that the method of proof used in the first lecture can be applied not only to ellipses, but to parabolas and hyperbolas as well, and that the closure theorem is true for all these curves. In fact, when Poncelet first worked with the closure theorem from the point of view of projective geometry, the properties of general conic sections (ellipse, parabola, hyperbola) were central in his mind.

In this section, we ask the question: "How many tangent lines can be drawn to a plane curve of degree 2 from a point not on the curve?" A projective transformation carries a line to a line and a plane curve of degree 2 to a plane curve of degree 2. It also carries a tangent line to a tangent line. Hence, by referring to the statement made at the end of the previous section, we know that there is a projective transformation that carries a plane curve of degree 2 to the unit circle. Therefore, it is sufficient to consider the case of the unit circle.

Note the following fact.

> Given the unit circle
>
> $$x_0^2 - x_1^2 - x_2^2 = 0 \tag{8}$$
>
> in the $\mathbb{P}^2(\mathbb{C})$, from each point P not on the unit circle one can draw two tangent lines to the circle.

PROOF. We prove the above theorem by considering three cases.

Case 1. From the point $(0 : 1 : m)$ on the line at infinity l_∞, but not on the unit circle (8), we can draw two lines tangent to the unit circle.

Since the point $(0 : 1 : m)$ does not lie on the unit circle (8), we have

$$1 + m^2 \neq 0. \tag{9}$$

As we mentioned in the previous section, the line passing through the point $(0 : 1 : m)$ is given by the equation

$$y = mx + l$$

in the xy-plane. By setting up the system of equations of the above line and of the unit circle in the xy-plane, and by eliminating the variable y, we obtain the discriminant of second degree. By equating the discriminant to 0 we get

$$(1 + m^2) - l^2 = 0.$$

When m and l satisfy this relation, the line becomes a tangent line. By setting

$$l = \pm\sqrt{1 + m^2}$$

and referring to expression (9), we obtain two tangent lines

$$y = mx \pm \sqrt{1 + x^2}$$

passing through the point $(0 : 1 : m)$.

Case 2. From the point $(0 : 0 : 1)$ on the line at infinity, we can draw two tangent lines.

The lines $x_0 \pm x_1 = 0$, that is, $\frac{x_1}{x_0} = \pm 1$, are tangent lines parallel to the y-axis that pass through the points $(1, 0)$ and $(-1, 0)$ on the unit circle viewed on the xy-plane.

Case 3. From the point (α, β) which is not on the unit circle on the xy-plane, we can draw two tangent lines.

If $\alpha = \beta = 0$, that is, if this point is the origin, we have shown that two tangent lines

$$y = \pm ix$$

can be drawn from the origin to the unit circle (8). Therefore, we consider the case where $(\alpha, \beta) \neq (0, 0)$. The unit circle (8) and the line at infinity l_∞ intersect at the points $(0, 1, \pm i)$. To determine whether two tangent lines to the unit circle passing through the point (α, β) exist, we need to analyze only the points on the unit circle that do not lie on l_∞. That is, it is sufficient to analyze the unit circle

$$x^2 + y^2 = 1 \tag{10}$$

in the xy-plane.

The equation for the tangent line passing through the point (a, b) on the circle (10) is given by

$$ax + by = 1.$$

(The equation for this tangent line is valid for complex numbers.) Therefore, to determine that there exist two tangent lines passing through the point (α, β), we need to find out whether there are two solutions to the system of equations

$$\begin{cases} a\alpha + b\beta = 1, \\ a^2 + b^2 = 1, \end{cases} \tag{11}$$

for a and b. Since the point (α, β) is not on the unit circle (10), we have the condition

$$\alpha^2 + \beta^2 \neq 1. \tag{12}$$

Since we have the condition $(\alpha, \beta) \neq (0, 0)$, we can consider the case where $\beta \neq 0$ (the proof for $\alpha \neq 0$ is similar).

From the first equation in system (11), we obtain

$$b = \frac{1}{\beta}(1 - a\alpha).$$

By substituting this expression for b in the second equation in (11) we obtain the following quadratic equation for a:

$$\left(1 + \frac{\alpha^2}{\beta^2}\right) a^2 - \frac{2\alpha}{\beta^2}\, a + \frac{1}{\beta^2} - 1 = 0.$$

(Referring to equation (12) we note that the coefficient of a^2 is not 0.) If we denote the discriminant of this quadratic equation by D, we have

$$\frac{D}{4} = \frac{1}{\beta^2}(\alpha^2 + \beta^2 - 1).$$

By equation (12) we know that $D \neq 0$. This, in turn, shows that this quadratic equation has two solutions. Hence, the system of equations has two distinct solutions. Thus we proved that there are two tangent lines passing through the point (α, β). This concludes the proof. □

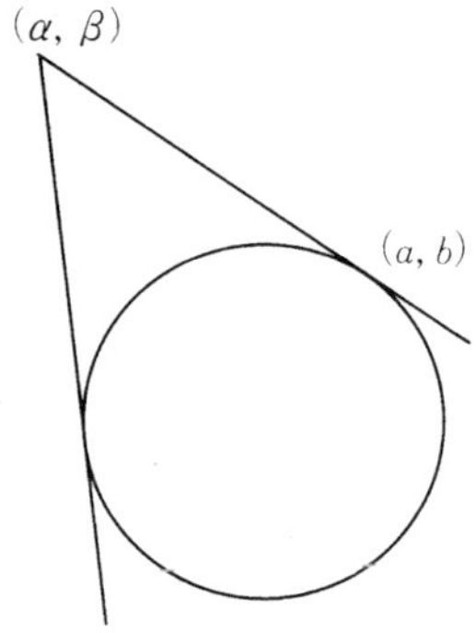

FIGURE 3.14

If we consider the unit circle in the world of real numbers, then from a point (α, β) we can draw two tangent lines to the unit circle whenever

$$\alpha^2 + \beta^2 > 1,$$

that is, whenever the point (α, β) lies outside of the circle. Also, we can draw two "complex" tangent lines whenever

$$\alpha^2 + \beta^2 < 1,$$

i.e., when the point (α, β) is inside the circle.

Summarizing, we can say that in $\mathbb{P}^2(\mathbb{C})$, for any plane curve of degree 2, we can draw two tangent lines to the curve from any given point not on the curve. Referring to the proof of the Poncelet theorem for the ellipse in Lecture 1, you will notice that the above fact demonstrates that $H(s_0, s_1)$ is a polynomial of second degree in s_0 and s_1. If we fix s_0 in the quadratic equation $H(s_0, s_1) = 0$, and the resulting equation in s_1 has two solutions in the set of real numbers, then it cannot have any other solutions.

3.5. Unit Circles and Lines in $\mathbb{P}^2(\mathbb{C})$

In the projective plane $\mathbb{P}^2(\mathbb{C})$, two distinct lines always intersect. In this sense, $\mathbb{P}^2(\mathbb{C})$ does not give the impression of an infinite object, which we feel when looking at $\mathbb{C}^2$. It might be rather convenient to view the projective plane as a plane having the finiteness property. The following example illustrates this property.

In $\mathbb{P}^2(\mathbb{C})$ a unit circle is identified with a line.

Any plane curve of degree 2 can be mapped to a unit circle by some projective transformation, and any line can be mapped to a line by another projective transformation. Therefore, the above fact can be restated as: **any plane curve of degree two is identified with a line.**

A word of caution before starting the proof. The complex coordinate plane $\mathbb{C}^2$ is actually four-dimensional. Since the coordinates x and y of a point (x, y) in $\mathbb{C}^2$ are two complex numbers, each of them varies in the complex plane (two-dimensional). Hence, the point (x, y) actually varies in four-dimensional space (see Chapter 2). Keeping this in mind, let us examine the line $y = 2x$ in $\mathbb{C}^2$. Since the independent variable x varies in the complex plane, this line is actually a two-dimensional surface in $\mathbb{C}^2$. Indeed, any line in $\mathbb{P}^2(\mathbb{C})$ appears as a sphere to our eyes (this will be demonstrated later).

Now we prove that the unit circle is identified with a line. First, we use a real plane to draw figures. Let P_0 be a point on the unit circle $x^2+y^2 = 1$. Draw a line l not passing through P_0. For a point P on the circle, draw the line P_0P and let Q be the point of intersection with l (see Figure 3.15). We obtain a correspondence

$$P \leftrightarrow Q$$

between P and Q.

In the above discussion, we have assumed that the line P_0P intersects the line l. This assumption indicated that our "construction" is performed in $\mathbb{P}^2(\mathbb{C})$. As P approaches P_0, the line P_0P gets closer to the tangent line to the unit circle at P_0. With P_0, we associate the point on the line l which is the intersection point of the tangent line at P_0 and the line l.

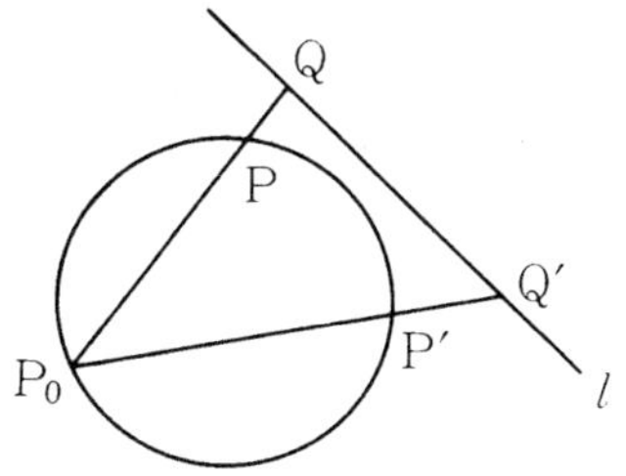

FIGURE 3.15

If we choose $(-1, 0)$ as the point P_0, and $x = 0$ for the line l, then the association between P and Q is represented by

$$P\left(\frac{1-t^2}{1+t^2}, \frac{2t}{1+t^2}\right) \leftrightarrow Q(0, t)$$

(see Figure 3.16).

In the case where the unit circle C is given by the equation

$$C\colon x_1^2 + x_2^2 - x_0^2 = 0 \tag{13}$$

in the projective plane $\mathbb{P}^2(\mathbb{C})$, the above correspondence is also valid. The point P_0 has coordinates $(1 : -1 : 0)$ and the line l is given by

$$x_1 = 0. \tag{14}$$

With each point P,

$$P(a_2^2 + a_1^2 : a_0^2 - a_1^2 : 2a_0a_1),$$

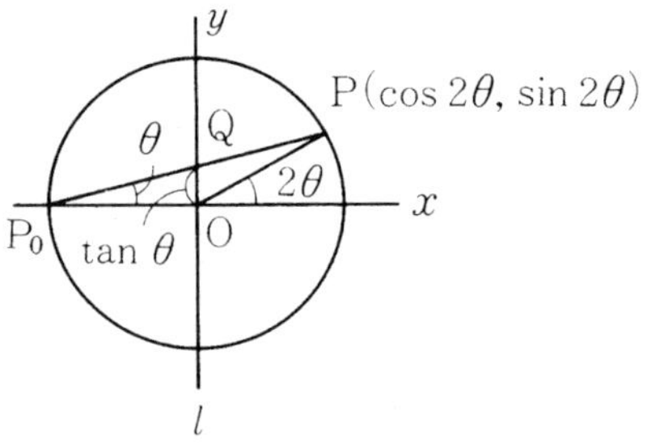

FIGURE 3.16. If $t = \tan\theta$, then $\cos 2\theta = \frac{1-t^2}{1+t^2}$ and $\sin 2\theta = \frac{2t}{1+t^2}$.

on C we associate the point

$$Q(a_0 : 0 : a_1)$$

on the line l (in fact, if $a_0 \neq 0$, by letting $t = \frac{a_1}{a_0}$, we obtain the previous correspondence). From this correspondence we know that the tangent line to the circle at P_0 is given by the equation $x = -1$, that is, by the equation $x_1 + x_0 = 0$. The point of intersection of the tangent line and the line $x = 0$ is the point at infinity $(0 : 0 : 1)$. Hence, we obtain the following correspondence:

$$P_0(1 : -1 : 0) \longleftrightarrow Q(0 : 0 : 1).$$

Thus we have shown that the unit circle C given by equation (13) and line l given by equation (14) are identified in $\mathbb{P}^2(\mathbb{C})$ (see Figure 3.17).

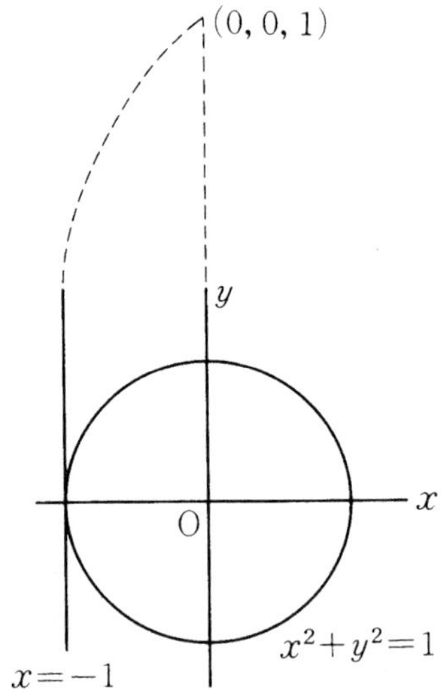

FIGURE 3.17

Since the line l is given by the equation $x_1 = 0$, it is the set of points of the form

$$(a_0 : 0 : a_1).$$

If $a_0 \neq 0$, by letting $t = \frac{a_1}{a_0}$, the point on the line l corresponding to t is given by

$$(a_0 : 0 : a_1) = (1 : 0 : t).$$

With this correspondence, the set of points on the line l such that $a_0 \neq 0$, is identified with the set of complex numbers denoted by $\mathbb{C}$. By adding the point at infinity

$$\infty = (0 : 0 : 1)$$

to the line l, we obtain the line given by equation (14). Therefore, we can write

$$l = \mathbb{C} \cup \infty.$$

The line l is a complex plane with a point at infinity adjoined; it is called the Riemann sphere. The relationship between the Riemann sphere and the complex plane is illustrated in Figure 3.18.

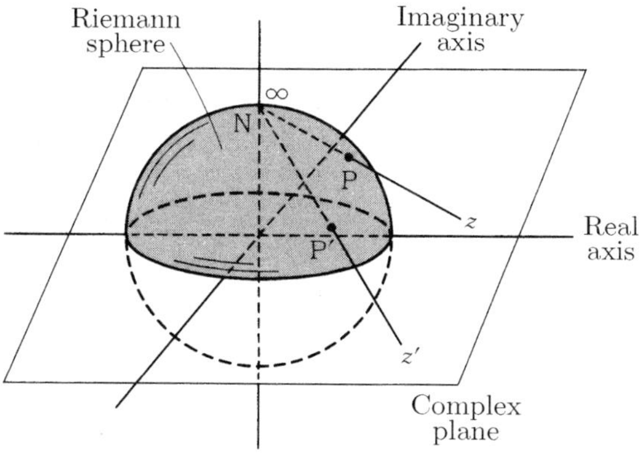

FIGURE 3.18. A point P on the sphere is identified with the point z in the complex plane $\mathbb{C}$ obtained from P by the projection from the north pole. The north pole represents the point at infinity.

In this manner we see that topologically, lines and plane curves of degree 2 in $\mathbb{P}^2(\mathbb{C})$ can be identified with a sphere (see Chapter 1).

However, there is yet more to the Poncelet theorem. The basic idea behind the discussion in Lecture 1 is as follows.

If we draw a tangent line from a point P on a plane curve C of degree 2 (we used an ellipse in Lecture 1) to an inner curve C' of degree 2, the process of determining the point P_1 at which the tangent line intersects the curve C again, is equivalent to solving the equation

$$H(s_0, s_1) = 0.$$

This equivalence shows that the totality of ordered pairs (P, P_1) of points P and P_1 on the curve C (regarded as a subset of $C \times C$) is a figure defined by an equation which is a quadratic equation in s_0 and in s_1. If we consider this figure in the complex plane, it is called an elliptic curve, and it is topologically equivalent to a torus. We discussed the torus in Chapter 2, but here the torus is given by an equation. The torus viewed as an elliptic curve has a much more complicated structure than a torus viewed topologically. Indeed, the notion of elliptic integrals came into being quite naturally as mathematicians were studying functions defined on elliptic curves. In elliptic integrals we see a connection between the algebraic proof in Lecture 1 and the analytic proof in Lecture 2. Algebra and analysis merged together through the geometric point of view on elliptic curves. In the merging of algebra and analysis we see the realization of the true harmony and beauty of mathematics. Regrettably, we cannot delve further into this merging of algebra and analysis, since it would require a great deal of additional preparation.

However, what we tried to convey to our readers is a feeling that when we look at an object, it might not be what it appears to be. It is our hope that through the study of the Poncelet theorem you have learned that an object which appears to be quite simple might actually have numerous facets. If such an object is observed from a higher mathematical "viewpoint", for instance, that of algebraic geometry, it becomes apparent that these facets are, in fact, unified.

We believe that mathematics has a timeless elegance. This sense of elegance continues to attract those who seek great personal satisfaction from this process of "climbing higher" in the hope of obtaining a more comprehensive point of view.

One last thought for our readers is that the science of mathematics is now, and will always be, in the process of evolution.

APPENDIX

Proof of the Poncelet Theorem Using Plane Geometry

We proved the Poncelet theorem for circles analytically in Lecture 2, using elliptic integrals. The reason elliptic integrals feature so prominently in Lecture 2 would become clear if we were using the approach based of the theory of algebraic functions. However, that theory lies beyond the scope of material discussed in this book. You might ask though, whether the theorem can be proved using plane geometry, at least if we limit ourselves to the case of circles. Indeed, since the properties of circles and lines are the main themes of plane geometry, the Poncelet theorem for circles can be proved geometrically. This geometric proof is not simple, but, fortunately, Ichiro Amemiya recently made our job a bit easier by introducing his relatively simple proof (though it still requires a great deal of work!). We present his proof here and hope that you will enjoy it.

A.1. Coaxal Circles

We will prove the Poncelet theorem for circles by means of plane geometry. We will use neither algebra nor analysis in this proof. Instead, we utilize profound properties of circles. We will prove each property we need as a separate proposition. It is customary to use the symbol O for the center of a circle O (in other words, O will denote both a circle and its center).

PROPOSITION 1. *Let O and O' be two circles with distinct centers O and O'. Let O' be inside O. Then the points such that the tangent segments*[1] *from each of them to the circles O and O' have equal length, form a line.*

PROOF. Let r and r' be the radii of the circles O and O'', respectively (Figure A.1). Let M be the point on the line passing through the centers O and O' such that the tangent segments from M to the circles O and O' have equal length. To find the point M, we first draw the lines through the centers O and O', respectively, that are perpendicular to the line OO' and choose the points A and A' on them such that the lengths of $\overline{OA}$ and $\overline{O'A'}$ are given by

$$\overline{OA} = r' \quad \text{and} \quad \overline{O'A'} = r,$$

[1]A tangent segment to a circle is part of a tangent line with one endpoint being the point of tangency.

respectively. Let M be the point of intersection of the line OO' and the perpendicular bisector of the segment AA'. Since

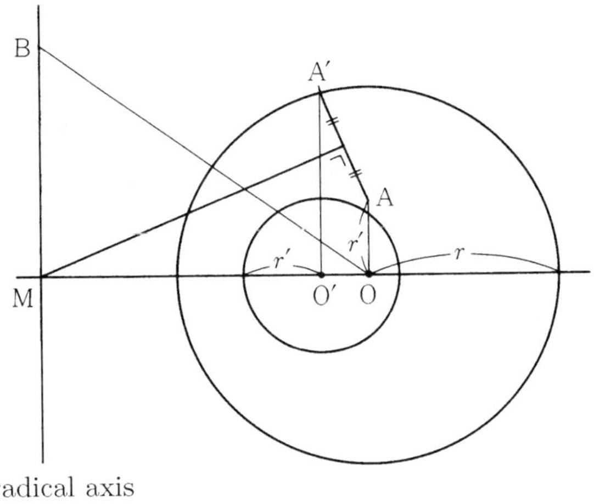

FIGURE A.1

$$\overline{MO}^2 + r'^2 = \overline{MA}^2 = \overline{MA'}^2 = \overline{MO'}^2 + r^2,$$

we have

$$\overline{MO}^2 - r^2 = \overline{MO'}^2 - r'^2.$$

The left and right sides of this equation are given by the square of the length of the tangent segments from M to O and O', respectively. (Verify this using the Pythagorean theorem.) Hence we found the required point M on the line OO'.

Let B be an arbitrary point on the line passing through M and orthogonal to the segment $\overline{MO}$. Then the square of the length of the tangent segment from B to O is equal to that of the tangent segment from B to O':

$$\begin{aligned}\overline{BO}^2 - r^2 &= \overline{BM}^2 + \overline{MO}^2 - r^2 = \overline{BM}^2 + \overline{MO'}^2 - r'^2 \\ &= \overline{BO'}^2 - r'^2.\end{aligned}$$

Hence, the line BM is the set of points described in Proposition 1. It is obvious that the lengths of tangent segments to O and O' from a point not on the line are not equal.

This concludes the proof. □

The line described in Proposition 1 is called the **radical axis** of the two circles O and O'.

Referring to Figure A.1 consider now the circle M centered at M of radius equal to the lengths of the tangent segments from M to O and O', respectively (the lengths are the same). Then the circle M is orthogonal to the circles O and O'.

♣ Two intersecting circles are called orthogonal if the tangent lines passing through the point of intersection are perpendicular (see Figure A.2). Stated otherwise, two intersecting circles are called orthogonal if the tangent line to one circle through the point of intersection passes through the center of the other circle (see Figure A.2).

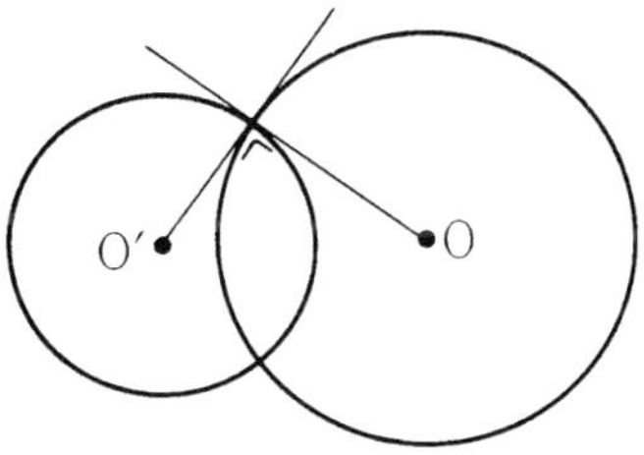

FIGURE A.2

Conversely, the fact that the circle M intersects O and O' orthogonally implies that the lengths of the tangent segments from M to O and O', respectively, are equal. Keeping this in mind, consider the circle O'' which is orthogonal to the circle M and has center located on the line MO on the same side of the circle O as the point M. The circle O'' then has the following property.

The radical axis of the circles O and O'' and that of O' and O'' are identical to the radical axis of O and O'.

The three circles O, O', and O'' with the relationship described above are called **coaxal** circles. Similarly, the circles O, O', O'', and $\widetilde{O}''$ are called **coaxal** circles if any two radical axes obtained from O, O', O'', and $\widetilde{O}''$ are identical (see Figure A.3).

PROPOSITION 2. *Let O, O' be two circles and P a point lying on the same side as O with respect to the radical axis of O and O'. Then there exists one and only one circle O'' passing through P and coaxal to the circles O and O'.*

PROOF. In the case where P is not on the circle M, let P' be the inverse of P in M, that is, P satisfies the equation

$$\overline{MP} \cdot \overline{MP'} = (\text{square of the radius of } M).$$

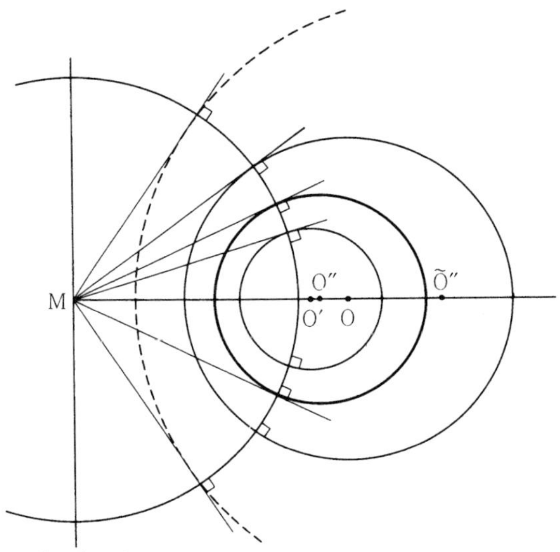

FIGURE A.3

Then, by the **Power Theorem**,[2] any circle passing through P and P' intersects the circle M orthogonally. Choose a circle O'' passing through P and P' and naving the center O'' on the line MO (the point O'' is the point of intersection of the line MO and the perpendicular bisector of the segment PP'). Then the circle O'' passes through the point P and is coaxal to the circles O and O' (see Figure A.4).

In the case where P is a point on M, we consider the circle O'' passing through P with center at the point of intersection of the line MO and the tangent line to M at P. The circle O'' passing through P at this intersection is coaxal to O and O'.

As a special case, if P is the intersection point N of the circle M and the segment MO, then the circle O'' of radius zero with center at N (that is, the circle O'' reduced to a point) is coaxal to O and O'.

---

[2]POWER THEOREM. *Given a circle O and a point outside O, draw a tangent segment joining P to the point of contact on O. Let T be the point of contact. Draw a secant segment from P intersecting the circle O at two points. Let A and B be the points of intersection. Then*

$$\overline{PA} \cdot \overline{PB} = \overline{PT}^2.$$

(See the figure below.)

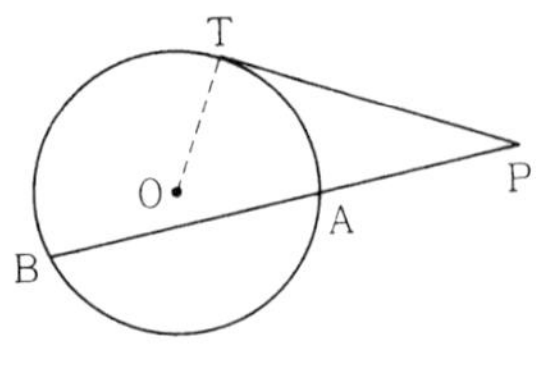

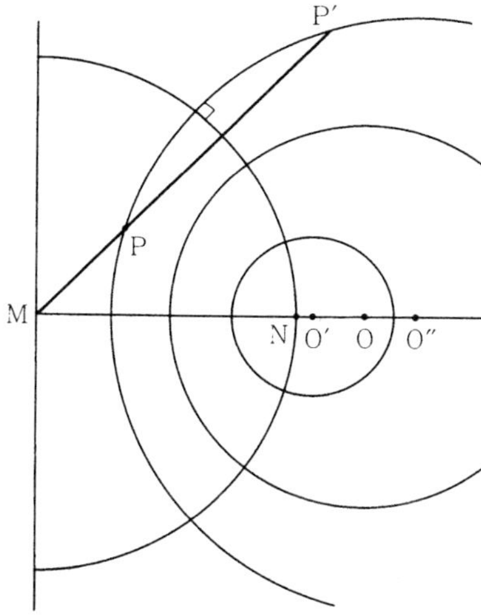

FIGURE A.4

It is clear from the construction that there exists only one such circle O''. This concludes the proof. □

PROPOSITION 3. *For a given chord CD of the circle O, there exists one and only one circle O'' which is tangent to CD and coaxal to O and O'.*

PROOF. In the case where the chord CD is parallel to the radical axis of O and O', consider a circle O'' (which we know exists by Proposition 2) coaxal to O and O', and passing through the point of intersection of CD and MO (see Figure A.5).

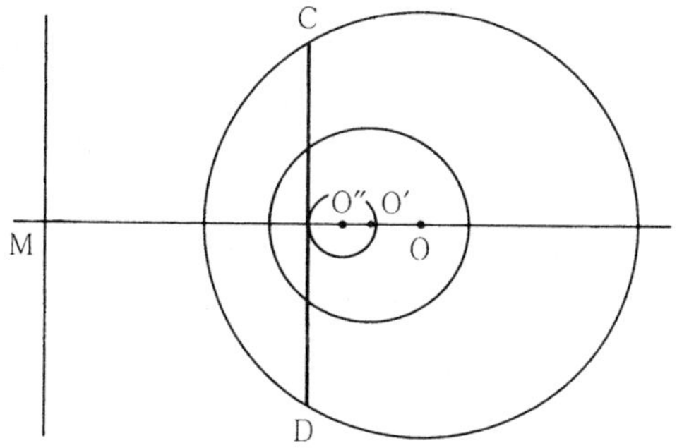

FIGURE A.5

In the case where the chord CD is not parallel to the radical axis of O and O', let B be the point of intersection of the line CD and the radical line. Take the point E on the chord CD such that the length of the segment BE is equal to the length of the tangent segment from B to the circle O (see Figure A.6). If we consider the circle O'' that passes through E and is coaxal

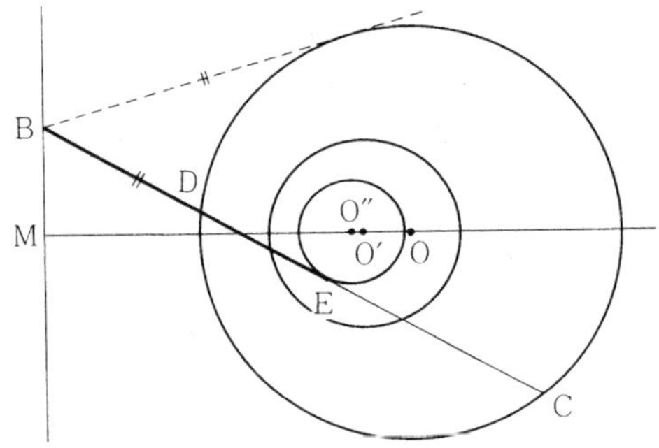

FIGURE A.6

to O and O'' (Proposition 2), then the length of the tangent segment from B to O'' is equal to the length of the tangent segment from B to O, which is equal to the length of BE. Therefore, the circle O'' is tangent to CD at E (see Figure A.6).

From this construction, we see that there is only one circle O'' tangent to CD and coaxal to O and O'.

This concludes the proof. □

PROPOSITION 4.

(i) *Given the circles O and O' as above, let O'' be a circle lying between O and O' and coaxal to O and O'. Then the ratio of the lengths of the tangent segments to O' and O'' from a point P on O is constant.*

(ii) *The circle O'' is uniquely determined by this constant ratio.*

PROOF. (i) We will use analytic geometry to prove Proposition 4. We first need to establish a coordinate system on the plane, so let us choose the line MO to be the x-axis, the radical axis to be the y-axis, and M to be the origin. Let the radius of the circle centered at M and orthogonal to the circles O, O', and O'' be 1. If the x-coordinates of the the centers of the circles O, O', and O'' are a, b, and c, respectively, then $a < b < c$. Furthermore, by the definition of the radical axis, the squares of the radii of the circles O, O', and O'' are given by

$$a^2 - 1, \quad b^2 - 1, \quad \text{and} \quad c^2 - 1,$$

respectively (see Figure A.7).

Let P be a point on O with coordinates (x, y). Then (x, y) satisfy the equation of the circle given by

$$(x - c)^2 + y^2 = c^2 - 1,$$

or equivalently,

(1) $$x^2 + y^2 + 1 = 2cx.$$

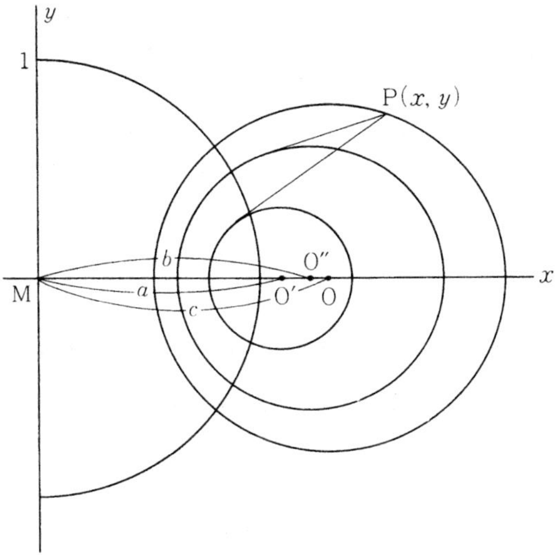

FIGURE A.7

Referring to Figure A.8, we see that the square of the length of the tangent segment to O' from P is given by

$$(x-a)^2+y^2-(a^2-1).$$

By substituting equation (1) for x^2+y^2+1 in the above expression and simplifying the result, we obtain the following:

(2) the square of the length of the tangent segment to O' from P equals $2(c-a)x$.

Similarly,

(3) the square of the length of the tangent segment to O'' from P equals $2(c-b)x$.

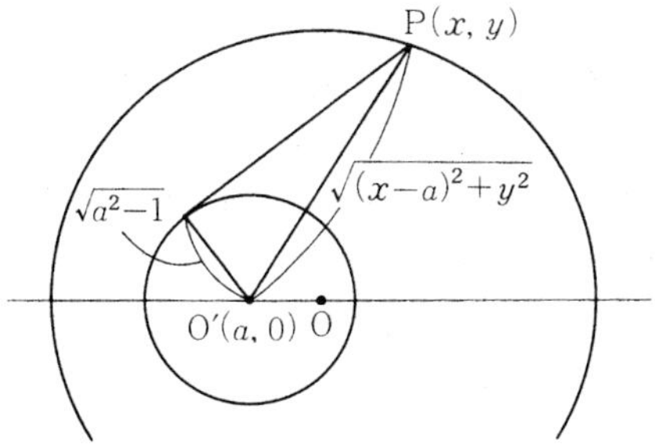

FIGURE A.8

The ratio of lengths in equations (2) and (3) is $(c-a)/(c-b)$. This ratio is constant and independent of the choice of the point $P(x, y)$. Thus, we have proved part (i) of Proposition 4.

(ii) Once we have a value of the ratio $(c-a)/(c-b)$, the value of b is uniquely determined by a and c. This implies that the circle O'' is uniquely determined by this ratio.

This concludes the proof. □

A.2. The Plane Geometry Proof of the Poncelet Theorem

The following property plays a significant role in the proof[3] of the Poncelet theorem.

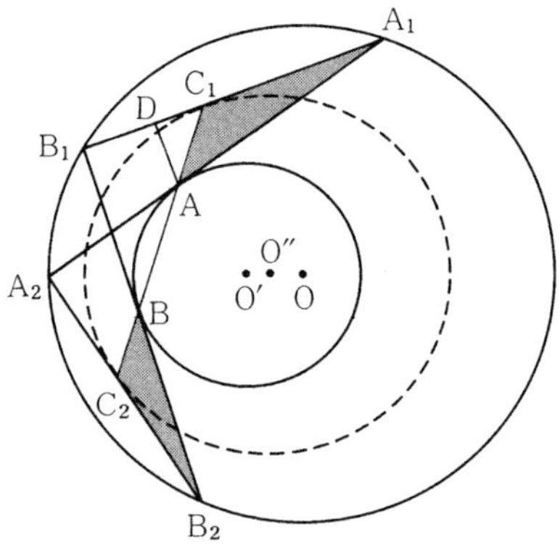

FIGURE A.9

PROPOSITION 5. *Suppose we have two circles O and O' with distinct centers O and O', respectively, where O' lies inside O. Let A_1A_2 and B_1B_2 be the chords of the circle O that are tangent to O' at A and B, respectively. Let B_1 lie on the opposite side from O' with respect to the chord A_1A_2. Then the circle O'' that is tangent to the chord A_1B_1 and coaxal to O and O', is also tangent to the chord A_2B_2.*

PROOF. Let C_1 and C_2 be the points of intersection of the line AB and the segments A_1B_1 and A_2B_2, respectively. Draw the line through A parallel to B_1B_2 and let D be the point of intersection of this line and the segment A_1B_1 (see Figure A.9). Then

$$\angle DAC_1 = \angle B_1BC_1 = \angle A_2AC_2 = \angle A_1AC_1.$$

[3]Credit for this proof goes to Ichiro Amemiya.

Hence, AC_1 is the angle bisector of $\angle DAA_1$. Therefore, we have

$$\begin{aligned}\overline{A_1A} : \overline{A_1C_1} &= \overline{DA} : \overline{DC_1} \quad \text{(property of an angle bisector of a triangle[4])}\\ &= \overline{B_1B} : \overline{B_1C_1} \quad (\triangle DAC_1 \sim \triangle B_1BC_1),\end{aligned}$$

that is,

$$\overline{A_1A} : \overline{B_1B} = \overline{A_1C_1} : \overline{B_1C_1}.$$

The left-hand side is the ratio of the lengths of the tangent segments to O' from A_1 and B_1, respectively. Hence, by Proposition 4, this ratio is equal to the ratio of the lengths of tangent segments to O''. Since we assumed that O'' is tangent to A_1B_1, we know that C_1 is the point of contact by referring to the right-hand side of the above equation.

Similarly, if we consider the circle O''' tangent to A_2B_2 and coaxal to O and O', then O''' is tangent to A_2B_2 at C_2.

On the other hand,

$$\angle C_1A_1A = \angle BB_2C_2 \quad \text{(inscribed angles, each intercepts the arc } B_1A_2 \text{ of } O\text{)},$$

$$\angle B_2BC_2 = \angle ABB_1 = \angle BAA_2 = \angle A_1AC_1.$$

Hence we have

$$\triangle A_1AC_1 \sim \triangle B_2BC_2.$$

From the above, we have

$$\overline{A_1A} : \overline{A_1C_1} = \overline{B_2B} : \overline{B_2C_2}.$$

By Proposition 4, a circle coaxal to O and O' is uniquely determined by the above ratio. Hence,

$$O'' = O'''.$$

This concludes the proof. □

We can now prove the Poncelet theorem as a consequence of Proposition 5.

---

[4]The property of the angle bisector of $\angle A$:

$$\triangle ACD \sim \triangle EBD,$$
$$AC : CD = BE : BD = AB : BD$$

(see the figure below).

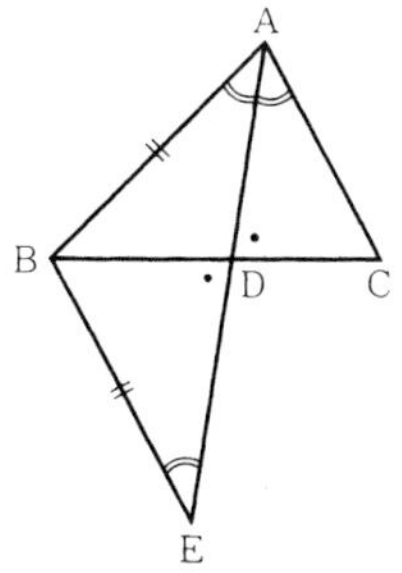

Proof of the Poncelet theorem. Since the proof of the Poncelet theorem for circles is essentially self-evident when the outer circle O and the inner circle O' are concentric, here we consider the case where O and O' are not concentric (that is, when the centers are distinct).

Referring to Figure A.9 (used in Proposition 5), we draw from the point A_2 a chord A_2A_3 of the circle O that is tangent to O'. Next, we draw from A_3 another chord A_3A_4 of the circle O that is tangent to O'. We continue this process of drawing the chords of O in succession until we obtain a series of chords A_1A_2, A_2A_3, ..., $A_{n-1}A_n$.

Similarly, starting from B_1, we construct a series of chords B_1B_2, B_2B_3, ..., $B_{n-1}B_n$ of O, each of which is tangent to O'.

By Proposition 5, each of the chords A_1B_1, A_2B_2, A_3B_3, ..., A_nB_n is tangent to O''. Hence, if the hypothesis of the Poncelet theorem is satisfied after n chords have been drawn, that is, if

$$A_1 = A_n,$$

then we can conclude that each of the chords A_1B_1 and $A_nB_n = A_1B_n$ is tangent to the same circle O''. Hence, B_1 must be identical to B_n, that is, $B_1 = B_n$.

Thus, we have proved the Poncelet theorem. □

So we see that, in the case of circles, there are two proofs distinct from the algebraic proof in Lecture 1 which involve the concept of "algebraic correspondence". The proof discussed in Lecture 2, based on Jacobi's ideas, introduces the invariant described by elliptic integrals. On the other hand, in Amemiya's proof, we encounter a circle coaxal to O and O' while successively constructing sequences of tangent lines to O' starting from points on the outer circle. These two methods of proof are entirely different from one another. And yet, if we focus on the idea of seeking something invariant in the process of constructing tangent lines, a common philosophy comes into focus.

Conclusion

A few years ago, one of our students came to visit us after attending some of our lectures. He said, "When I was in high school, I thought mathematics was something utterly unnatural. This may have been because of the textbooks we used. However, after attending your lectures I began to get a feeling that mathematics is, indeed, rather interesting!" The concept of mathematics being something "unnatural" had not entered our minds before it was so described by our student. But now we think that this description succinctly expresses how high school students often perceive mathematics as something stiff and formal.

I believe mathematics is a malleable and magnanimous discipline. When we encounter a mathematical problem, we ask ourselves "Why is it so?" and, we discover that mathematics gives us an answer in various forms. At times the answer is simple and intuitive. Other times, one arrives at the answer to a question only after a great deal of effort. It is deeply layered, and displays, upon further inquiry, additional facets not previously revealed.

The study of geometric figures from a global point of view (topology) has revealed characteristic properties of these figures which could have never been derived from or understood by classical geometry. The investigation of the simple question "What is dimension?" deepened our understanding of the space and time in which we live. This same, seemingly simple, question about dimension caused mathematicians and physicists to ponder upon the mysterious realms of higher dimensions. As we discussed in Chapter 3, an insight into the history of mathematics gave us an understanding of the concept of infinity. Mathematicians began to understand the concept of infinity even before a formal introduction of calculus, as they attempted to clarify the "light filtering in from infinity." When we try to shed light upon the mystery of the Poncelet theorem, we see that algebra is interwoven with geometric figures. This interweaving leads us, in turn, to the study of algebraic geometry involving the system of complex numbers.

Mathematicians are always led on by the siren song of mathematical questions. Like Indiana Jones in his quest for the golden chalice, mathematicians, in their search for answers, are driven into deeper and deeper levels of investigation. Hopefully, in our lectures we have shown you a different facet of mathematics that has triggered in you a desire for further investigation into mathematics mysteries. You may have felt some initial hesitation to take this further step into the deeper study of mathematics because you feel that it might prove to be too difficult. However, we believe studying a discipline such as mathematics is like a journey. Whether you merely glance at the scenery passing by or stop to gaze at it more intently, is up to you.

We believe that a lecture is an effective format for encouraging the exchange of ideas (as opposed to a dry recitation of facts). It is even better to finish the lecture as an open forum in the hope that it will encourage students to begin their own investigation into those areas of mathematics they find of special interest.

This is the end of our lectures. We hope we have led you to the spring of mathematical knowledge, and that you will drink in it deeply.

We thank Mr. Hisao Miyauchi of Iwanami Shoten for his effort in the composition and preparation of the illustrations for this book. We hope that our readers find it to be both useful and inspirational. Thank you.

Titles in This Series

20 **Kenji Ueno, Koji Shiga, and Shigeyuki Morita,** A mathematical gift, II: The interplay between topology, functions, geometry, and algebra, 2004

19 **Kenji Ueno, Koji Shiga, and Shigeyuki Morita,** A mathematical gift, I: The interplay between topology, functions, geometry, and algebra, 2003

18 **Timothy G. Feeman,** Portraits of the Earth: A mathematician looks at maps, 2001

17 **Serge Tabachnikov, Editor,** Kvant Selecta: Combinatorics, I, 2001

16 **V. V. Prasolov,** Essays on number and figures, 2000

15 **Serge Tabachnikov, Editor,** Kvant Selecta: Algebra and analysis. II, 1999

14 **Serge Tabachnikov, Editor,** Kvant Selecta: Algebra and analysis. I, 1999

13 **Saul Stahl,** A gentle introduction to game theory, 1999

12 **V. S. Varadarajan,** Algebra in ancient and modern times, 1998

11 **Kunihiko Kodaira, Editor,** Basic analysis: Japanese grade 11, 1996

10 **Kunihiko Kodaira, Editor,** Algebra and geometry: Japanese grade 11, 1996

9 **Kunihiko Kodaira, Editor,** Mathematics 2: Japanese grade 11, 1997

8 **Kunihiko Kodaira, Editor,** Mathematics 1: Japanese grade 10, 1996

7 **Dmitry Fomin, Sergey Genkin, and Ilia Itenberg,** Mathematical circles (Russian experience), 1996

6 **David W. Farmer and Theodore B. Stanford,** Knots and surfaces: A guide to discovering mathematics, 1996

5 **David W. Farmer,** Groups and symmetry: A guide to discovering mathematics, 1996

4 **V. V. Prasolov,** Intuitive topology, 1995

3 **L. E. Sadovskiĭ and A. L. Sadovskiĭ,** Mathematics and sports, 1993

2 **Yu. A. Shashkin,** Fixed points, 1991

1 **V. M. Tikhomirov,** Stories about maxima and minima, 1990